中国世界级非遗文化悦读系列·寻语识遗
丛书主编 魏向清 刘润泽

中国雕版印刷技艺

（汉英对照）

刘韶方 单旭光 主编

Chinese Engraved Block Printing

南京大学出版社

本书为以下项目的部分成果：

南京大学外国语学院"双一流"学科建设项目

全国科学技术名词审定委员会重点项目"中国世界级非物质文化遗产术语英译及其译名规范化建设研究"

教育部学位中心 2022 年主题案例项目"术语识遗：基于术语多模态翻译的中国非物质文化遗产对外译介与国际传播"

南京大学 – 江苏省人民政府外事办公室对外话语创新研究基地项目

江苏省社科基金青年项目"江苏世界级非物质文化遗产术语翻译现状与优化策略研究"（19YYC008）

江苏省社科基金青年项目"江苏世界级非遗多模态双语术语库构建研究"（23YYC008）

南京大学暑期社会实践校级特别项目"讲好中国非遗故事"校园文化活动

参与人员名单

丛书主编 魏向清　刘润泽
主　　编 刘韶方　单旭光
翻　　译 张　璐　李　廉
译　　校 Zhujun Shu　Benjamin Zwolinski
学术顾问 李江民　李华俊
出版顾问 何　宁　高　方
中文审读专家（按姓氏拼音首字母排序）
　　　　　　陈　俐　丁芳芳　王笑施
英文审读专家 Colin Mackerras　Leong Liew
参编人员（按姓氏拼音首字母排序）
　　　　陈红利　江　娜　梁鹏程　秦　曦　孙文龙　王　倩
　　　　武志琳　许沁杨　杨艳霞　张文奕　朱云涛
手　　绘 乌兰托雅
实 物 图 扬州江民雕版印刷有限公司
知识图谱 王朝政
中国历史纪年简表 王朝政
特别鸣谢 江苏省非物质文化遗产保护研究所
　　　　　扬州江民雕版印刷有限公司

编者前言

2019年秋天开启的这次"寻语识遗"之旅,我们师生同行,一路接力,终于抵达了第一个目的地。光阴荏苒,我们的初心、探索与坚持成为这5年奔忙的旅途中很特别,也很美好的回忆。回望这次旅程,所有的困难和克服困难的努力,如今都已经成为沿途最难忘的风景。

这期间,我们经历了前所未有的自主性文化传承的种种磨砺,创作与编译团队的坚韧与执着非同寻常。古人云,"唯其艰难,方显勇毅;唯其磨砺,始得玉成"。现在即将呈现给读者的是汉英双语对照版《中国世界级非遗文化悦读系列·寻语识遗》丛书(共10册)和中文版《中国世界级非遗文化悦读》(1册)。书中汇聚了江苏牵头申报的10项中国世界级非物质文化遗产项目内容,我们首次采用"术语"这一独特的认知线索,以对话体形式讲述中国非遗故事,更活泼生动地去诠释令我们无比自豪的中华非遗文化。

2003年,联合国教科文组织(UNESCO)第32届会议正式通过了《保护非物质文化遗产公约》(以下简称《公约》),人类非物质文化遗产保护与传承进入了全新的历史时期。20多年来,世界"文化多样性"和"人类创造力"得到前所未有的重视和保护。截至2023年12月，中国被列入《人类非物质文化遗产代表作名录》的项目数量位居世界之首（共43项），是名副其实的世界非遗大国。正如《公约》的主旨所述，非物质文化遗产是"文化多样性保护的熔炉，又是可持续发展的保证"，中国非遗文化的世界分享与国际传播将为人类文化多样性注入强大的精神动力和丰富的实践内容。事实上，我国自古就重视非物质文化遗产的保护与传承。"收百世之阙文，采千载之遗韵"，现今留存下来的卷帙浩繁的文化典籍便是记录和传承非物质文化遗产的重要载体。进入21世纪以来，中国政府以"昆曲"申遗为开端，拉开了非遗文化国际传播的大幕，中国非遗保护与传承进入国际化发展新阶段。各级政府部门、学界和业界等多方的积极努力得到了国际社会的高度认可，中国非遗文化正全面走向世界。然而，值得关注的是，虽然目前中国世界级非物质文化遗产的对外译介与国际传播实践非常活跃，但在译介理据与传播模式方面的创新意识有待加强，中国非遗文化的国际"传播力"仍有待进一步提升。

《中国世界级非遗文化悦读系列·寻语识遗》这套汉英双语丛书的编译就是我们为中国非遗文化走向世界所做的一次创新译介努力。该编译项目的缘起是南京大学翻译专业硕士教育中心特色课程"术语翻译"的教学实践与中国文化外译人才培养目标计划。我们秉持"以做促学"和"全过程培养"的教学理念，探索国别化高层次翻译专业人才培养的译者术语翻译能力提升模式，尝试走一条"教、学、研、产"相结合的翻译创新育人之路。从课堂的知识传授、学习，课后的合作研究，到翻译作品的最终产出，我们的教研创新探索结出了第一批果实。

汉英双语对照版丛书《中国世界级非遗文化悦读系列·寻语识遗》被列入江苏省"十四五"时期重点图书出版规划项目，这是对我们编译工作的莫大鼓励和鞭策。与此同时，我们受到来自国际中文教育领域多位专家顾问的启发与鼓励，又将丛书10册书的中文内容合并编成了一个合集《中国世界级非遗文化悦读》，旨在面向国际中文教育的广大师生。2023年夏天，我们这本合集的内容经教育部中外语言交流合作中心教研项目课堂试用，得到了非常积极的反馈。这使我们对将《中国世界级非遗文化悦读》用作非遗文化教材增添了信心。当然，这个中文合集版本也同样适用于国内青少年的非遗文化普及，能让他们在"悦读"过程中感受非遗文化的独特魅力。

汉英双语对照版丛书的编译理念是通过"术语"这一独特认知路径，以对话体形式编写术语故事脚本，带领读者去开启一个个"寻语识遗"的旅程。在每一段旅程中，读者可跟随故事里的主人公，循着非遗知识体系中核心术语的认知线索，去发现、去感受、去学习非遗的基本知识。这样的方式，既保留了非遗的本"真"知识，也彰显了非遗的至"善"取向，更能体现非遗的大"美"有形，是有助于深度理解中国非遗文化的一条新路。为了让读者更好地领会非遗知识之"真善美"，我们将通过二维码链接到"术语与翻译跨学科研究"公众号,计划陆续为所有的故事脚本提供汉语和英语朗读的音频,并附上由翻译硕士专业同学原创的英文短视频内容,逐步完成该丛书配套的多模态翻译传播内容。这其中更值得一提的是,我们已经为这套书配上了师生原创手绘的核心术语插图。这些非常独特的用心制作融入了当代中国青年对于中华优秀传统文化的理解与热爱。这些多模态呈现的内容与活泼的文字一起将术语承载的厚重知识内涵,以更加生动有趣的方式展现在读者面前,以更加"可爱"的方式讲好中国非遗故事。

早在10多年前,全国高校就响应北京大学发起的"非遗进校园"倡议,成立了各类非遗文化社团,并开展了很多有益的活动,初步提升了高校学生非遗文化学习的自觉意识。然而,我们发现,高校学生群体的非遗文化普及活动往往缺乏应有的知识深度,多限于一些浅层的体验性认知,远未达到文化自知的更高要求。我们所做的一项有关端午非遗文化的高校学生群体调研发现,大部分高校学生对于端午民俗的了解较为粗浅,相关非遗知识很是缺乏。试问,如果中国非遗文化不能"传下去",又怎能"走出去"?而且,从根本上来说,没有对自身文化的充分认知,是谈不上文化自信的。"求木之长者,必固其根本;欲流之远者,必浚其泉源。"中国世界级非遗文化的对外译介与国际传播要解决的关键问题是培养国人尤其是青少年的非遗文化自知,形成真正意义上基于文化自知的文化自信,然后才有条件由内而外，加强非遗文化的对外译介与国际传播。非遗文化小书的创新编译过程正是南京大学"非遗进课堂"实践创新的成果，也是南大翻译学子学以致用、培养文化自信的过程。相信他们与老师一起探索与发现，创新与传承，译介与传播的"寻语识遗"之旅定会成为他们求学过程中一个重要的精神印记。

我们要感谢为这10个非遗项目提供专业支持的非遗研究与实践方面的专家，他们不仅给我们专业知识方面的指导和把关，而且也深深影响和激励着我们，一步一个脚印，探索出一条中国非遗文化"走出去"和"走进去"的译介之路。事实上，这次非常特别的"寻语识遗"之旅，正是因为有了越来越多的同行者而变得更加充满希望。最后，还要特别感谢南京大学外国语学院给了我们重要的出版支持，特别感谢所有参与其中的青年才俊，是他们的创意和智慧赋予了"寻语识遗"之旅始终向前的不竭动力。非遗文化悦读系列是一个开放的非遗译介实践成果系列，愿我们所开辟的这条"以译促知、以译传通"的中国非遗知识世界分享的实践之路上有越来越多的同路人，大家携手，一起为"全球文明倡议"的具体实施贡献更多的智慧与力量。

目　录
Contents

百字说明　A Brief Introduction

内容提要　Synopsis

知识图谱　Key Terms

《金刚经》 *The Diamond Sutra* ················· 001

雕版印刷　Engraved Block Printing ················· 010

制版　Block Making ················· 029

松烟墨　Pine-Soot Ink ················· 040

连史纸　*Lianshi* Paper ················· 049

写版与校正　Sampling and Proofreading ················· 062

上样　Sample Pasting ················· 073

刻版　Block Engraving ················· 079

刷印　Brush Printing ················· 091

线装　Thread Binding ················· 102

套色印刷　Overprinting ················· 112

饾版和拱花　Block-Assembled Overprinting and Embossed Overprinting ················· 123

刻书　Block-Printed Books ……………………………………138

套色版画　Overprinted Woodcuts ………………………………147

结束语　Summary ………………………………………………163

中国历史纪年简表　A Brief Chronology of Chinese History ………165

百字说明

　　雕版印刷术是中国古代重要发明之一，是世界现代印刷术最古老的源头。雕版印刷融合了造纸、制墨、雕刻、摹拓等多种中国传统技术和工艺，将文字和图像反向雕刻在木板上，制成印版，然后在印版上刷墨、铺纸，再将图文转印到纸张上。2009年，雕版印刷术被联合国教科文组织列入《人类非物质文化遗产代表作名录》。

A Brief Introduction

Chinese Engraved Block Printing is considered the earliest origin of modern printing. As one of the important inventions in ancient China, it incorporates various traditional Chinese techniques and crafts, including paper-making, ink-making, engraving, and rubbing. Before printing words or images on a piece of paper, characters and pictures are first reversely engraved on wooden boards, known as printing blocks, to which the ink is then applied, and texts and images are transferred finally onto the paper. This craft was inscribed on the UNESCO's Representative List of the Intangible Cultural Heritage of Humanity in 2009.

内容提要

为全面认识中国古老的雕版印刷技艺，小龙和大卫专门去请教雕版印刷史专家张教授，初步了解了雕版印刷术的发展历程。随后，他们前往扬州中国雕版印刷博物馆，实地考察了雕版印刷的主要工艺流程和所用材料。最后，在桃花坞木刻年画社，小龙和大卫还亲自体验了木刻年画的刷印过程。

Synopsis

Xiaolong and David paid a special visit to Prof. Zhang, an expert on Chinese engraved block printing. They gained a basic understanding of its development from his introduction. They then explored the entire process of engraved block printing at China Engraved Block Printing Museum in Yangzhou. Finally, they learned how to create New Year pictures at Taohuawu Woodcut New Year Picture Society in Suzhou.

知识图谱
Key Terms

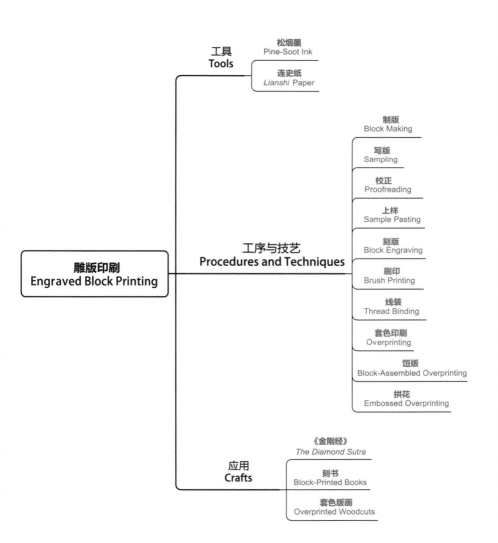

《金刚经》

> 小龙和大卫前去拜访张教授，他是研究中国雕版印刷史的著名专家。

小　　龙：张教授早上好！这是我的英国朋友大卫。谢谢您抽空接待我们。

张教授：你们好，欢迎欢迎，难得你们年轻人对中国雕版印刷感兴趣。

小　　龙：张教授，我们想多了解一些这方面的知识。

张教授：好啊，你们对哪些方面比较感兴趣呢？

小　　龙：我查了一些资料，关于雕版印刷术的年代，最多的说法是唐朝已经开始了规模印刷。现存最早的雕版印刷品是唐朝的吗？

张教授：是的，是唐朝印制的《金刚经》卷轴，全名叫《金刚般若波罗蜜经》。它是世界上现存最早、标有刻

《金刚经》 *The Diamond Sutra*

印日期的雕版印刷品。准确地说,这是目前有记录的最早的雕版印刷书籍。

小　龙：是在哪里发现的？

张教授：在甘肃敦煌莫高窟,但应该是四川刻印的。卷尾上印有"咸通九年四月十五日王玠为二亲敬造普施"这十八个字。这一行字传递的信息对雕版印刷历史的研究特别重要。

大　　卫：咸通九年是哪一年？这句话讲的是什么呢？

张教授：是公元868年，唐朝末年，当时正是佛教盛行的时候。人们常用抄经、刻经的方法积德行善。这句话的意思是王玠抄写经文，请人刻印，然后免费散发，以此为父母祈福。

大　　卫：原来是这样。那要印很多份了。

张教授：没错。这也说明当时雕版印刷很盛行。

小　　龙：这部《金刚经》现藏哪里？

张教授：伦敦大英图书馆。稍等，我这里刚好有这个印本的图片。你们看，全卷由七张纸连接而成，长491.5厘米，宽30.5厘米。不过很可惜，在英国修复时，整卷纸被裁成几段，放在了一块块独立的展板上。

小　　龙：将近5米，很长呀。不过，这样分割开，就不像一本完整的书了。

张教授：是的，这完全不是中国卷轴的展现方式。

大　　卫：张教授，这本《金刚经》为什么会保存在大英图书馆呢？

张教授：1907年，英国人斯坦因率领考察队到中国。他们从敦煌把大量的莫高窟文物运回英国，其中就有这本《金刚经》。后来《金刚经》被大英博物馆收藏，再后来又转到大英图书馆保存。

大　卫：这么多年了，上面的字迹还是很清晰的。

张教授：是啊，这部《金刚经》无论是第一页的木刻版画，还是后面的经文，印刷都很精美。版画上有19个人物，经文部分每行19个字，一共5000多字。你们看，它构图复杂，刻工精湛，墨色均匀，字迹清晰。

小　龙：太珍贵了！看来当时的雕版印刷技术已经很成熟了。

张教授：从文献上看，雕版印刷的书主要是供初学者识字，印刷的其他物品供人们日常使用。当然了，道士和佛教徒也会利用这种新技术普及宗教知识，弘扬教义。

The Diamond Sutra

> Xiaolong and David visited Prof. Zhang, a well-known expert on the history of Chinese engraved block printing.

Xiaolong: Good morning, Prof. Zhang. Thank you for making time to meet us. This is my friend David from England.

Prof. Zhang: Nice to meet you. I'm glad to know you are interested in Chinese engraved block printing. You are more than welcome!

Xiaolong: Thank you. We wish to learn more about the technique.

Prof. Zhang: Good. Is there anything specific you are interested in?

Xiaolong: I did a bit of research. It says that the Tang Dynasty was the time when this kind of printing was done on a large scale. Does that mean it's the earliest time when the block-printed works were found?

Prof. Zhang: Yes. One of the most famous works is the scroll of *The Diamond Sutra*. Its full name is *The Diamond Perfection of Wisdom Sutra*. It's the earliest block-printed book in the world with an exact date printed on it.

Xiaolong: So, it's more than 1,400 years old. Where was it discovered?

Prof. Zhang: At the Mogao Grottoes in Dunhuang, Gansu Province. And it was probably engraved and printed in Sichuan Province. Its end note says "Respectfully made for his parents by Wang Jie on April 15th of the 9th year of Xiantong". The sentence contains information that is highly significant for researching the history of engraved block printing.

David: When was that? And what was it about?

Prof. Zhang: It was 868 A.D., at the end of the Tang Dynasty.

At that time, Buddhism was flourishing in China. It was popular for people to transcribe or engrave Buddhist scriptures to seek blessings for their families. The note says Wang Jie transcribed the scriptures himself devoutly, paid for the printing and distributed their copies for free. By doing this, he hoped the Buddha would bless his parents.

David: I see. So, he must have had many copies printed.

Prof. Zhang: Sure. It suggested that the technique was very popular then.

Xiaolong: Where is the scroll kept now?

Prof. Zhang: It's in the British Library, located in London. Hang on a minute, I've got a photocopy of the original. You see, it should have been a scroll of 7 pieces of paper, about 16 feet long and 1 foot wide. Unfortunately, the scroll was cut into several pieces and displayed on separate boards after its restoration in the UK.

Xiaolong: Nearly 16 feet? That's fairly long. But it's not like a complete book at all if they are cut separately.

Prof. Zhang: You're right. It's definitely not the right way to display a Chinese scroll.

David: Prof. Zhang, why is it kept in the British Library?

Prof. Zhang: It's a long story. In 1907, Marc Stein, an Englishman, led an expedition to Dunhuang and took large quantities of cultural relics from the Mogao Grottoes back to the UK, including this scroll. It was stored first in the British Museum and then transferred to the British Library.

David: The characters on the scroll are still legible after so many years.

Prof. Zhang: Yes, the printing is very beautiful. There are 19 figures in the painting on the first page and more than 5,000 characters in the scripture, each line containing 19 characters. It's famous for its complicated structure, exquisite engraving, clear ink colour, and handsome handwriting.

Xiaolong: Very precious indeed! It shows an advanced printing technique in China at that time.

Prof. Zhang: Right. It's recorded that the block-printed books were mainly for beginners to learn how to read and other block-printed materials for people's daily use. What's more, Taoist and Buddhist monks also took advantage of it to introduce their beliefs and doctrines.

雕版印刷

小　　龙：张教授，您刚才谈到雕版印刷为文化传播创造了更好的条件。这项技艺是不是很复杂呀？

张教授：是的，这项技术的前期工序很繁琐费力，但后期印刷阶段就可以批量生产了。

大　　卫：雕版印刷都有哪些工序呢？

张教授：工序还不少呢。第一步抄写文稿，第二步将稿纸反贴在木板上，形成反体字，第三步雕刻文字，用刻刀把版面上的字刻成阳文。之后，还要在凸起字体上刷墨，用纸覆压刷印出来，最后装订成册。

大　　卫：还挺复杂。

小　　龙：雕版印刷的步骤确实挺多。不过，一次可以印制很多本，比原先抄写的效率高多了。这样既解放了生产力，又提高了文化传播速度。

张教授：小龙总结得很好，雕版印刷很了不起。

小　　龙：好像古书的宋刻本特别出名，这是为什么呢？

张教授：这主要是因为宋朝经济和技术都很发达，文学艺术也很繁荣，所以书籍需求量大。

大　　卫：宋朝印刷需求量大，所以宋刻本很有名，是吗？

张教授：说得有道理，但宋刻本出名，还因为一个关键技术。宋代之前没有排版一说，刻版比较随意。到了宋朝，雕版印刷由政府监管，主要由国子监负责。国子监对工序、方法进行了标准化、规范化管理，进一步完善了雕版印刷技术，所以雕版技艺更精湛，印书质量更好。

小　　龙：难怪都说宋刻本是研究雕版印刷最好的样本。但这项传统技艺的成熟应该也经历了很长时间吧？

张教授：是的，要是从头说起，时间的确很漫长。它先后经历了印章、墨拓石碑、木拓法帖等几个阶段，最终才出现了成熟的雕版印刷技术。

小　　龙：张教授，印章最早出现在哪个朝代？

张教授：根据出土文物和历史记载，2700多年前的东周时期，就出现了印章。起初，印章只是做生意用的凭证，后来成为代表权力的法物。

小　　龙：嗯，我们现在还用印章呢。印章应该是中国人特有的一种签名方式吧。

大　卫：我们不一样，一般都是直接签名。

张教授：中外确实不太一样。而且在中国，印章对书法的发展有过积极影响。2000多年前的汉代，人们受印章启发，开始用阴文篆刻的方法把一些著名的书法作品刻在石碑上，让人们学习临摹，后来又进一步把整部书刻到石碑上。

小　龙：然后再从石碑上把书法作品拓下来？

张教授：是的。东晋时期，为了让更多的人欣赏和临摹石碑上的内容，出现了石碑拓印技艺。就是把纸铺在石碑上，用墨包捶打纸张，把碑上的字拓印下来。

小　龙：这个我知道。今天练书法用的法帖大多是拓印的。那木拓法帖是什么？是不是把文字刻在木板上呀？

张教授：是的。相比石碑拓印，木拓法帖是一种更接近雕版印刷的技艺。这需要将纸铺在刻好文字和图案的木版上，用墨包捶打纸张，把木板上的内容拓印到纸上。

大　卫：木板上刻字肯定要容易很多。

张教授：那当然了。另外，还有一种捺印技艺，对雕版印刷也有一定的启发。捺印来自印度佛教，主要用于佛像印制。捺印也是在木板上刻字、刻图案，不同之

雕版印刷　Engraved Block Printing　013

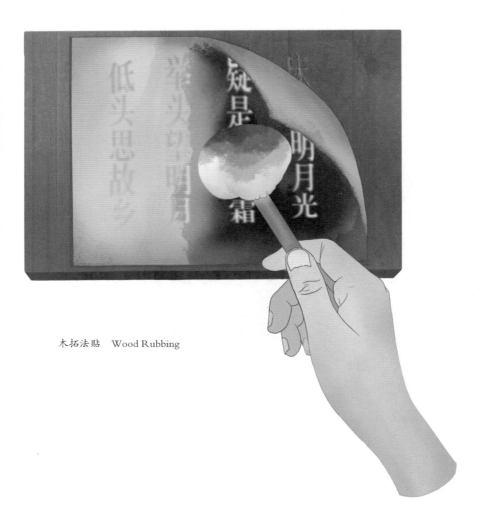

木拓法贴　Wood Rubbing

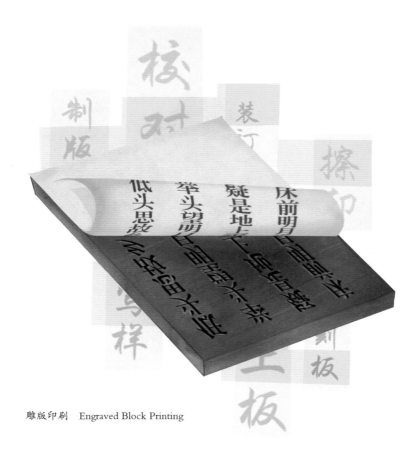

雕版印刷　Engraved Block Printing

处是将印版蘸上印泥，向下压印到纸面上。

大　卫：感觉像是盖了个佛像戳。

张教授：说得很形象。

小　龙：然后就发展到雕版印刷了吧？

张教授：是的，从石碑拓印到木拓法帖，后来就出现了雕版印刷。再后来，又出现了更便捷的活字印刷。具体说，就是事先准备好足够的单个活字，随时拼版，可大大缩短制版时间。

小　龙：活字印刷应该是宋朝出现的吧？

张教授：是的。宋朝商业非常发达，技术创新不断。纸币、指南针、风力水车、纺织机等和工业有关的技术发明都得到了应用。宋朝学者第一次详细记录了这些新技术、新事物。而这一切能流传至今，又得益于规模印刷。

小　龙：可活字印刷发明之后并没有流行起来，这又是什么原因呢？

张教授：原因有很多。第一是技术上的原因。你们知道，时间、环境等因素会造成泥土和木材质地一定的变化。如果做成一个个独立的字，印刷过程中的伸缩率不一样，可能会出现高低不平、大小不一的情况，所以很难确保印刷品质量的稳定一致。

小　　龙：明白了，活字印刷没有流行是因为活字的材质问题。

张教授：是的。第二个原因是，虽然当时印刷品种类很多，但古代知识更新慢，印制的书目相对固定。主要的印刷品是四书五经①类的儒家经典，以及一些佛教、道教的书。还有年画②、信笺③、账本和历书等日常印刷品，但这些内容也相对固定，不需要每次印刷都重新拼版。

大　　卫：我明白了。当时印刷制品的内容比较固定，显现不出活字印刷的优势。

张教授：没错。而且当时大多数印刷匠人不识字，所以活字印刷排版很困难。

小　　龙：那就是说，当时的活字印刷缺少规模发展的技术动力和社会需求。

张教授：非常正确。此外，雕版印刷在中国出现后，先后传到朝鲜和日本，又通过丝绸之路，传到了中亚、西亚和欧洲。可以说，中国雕版印刷不仅历史悠久，而且国际影响也非常广泛，曾被誉为印刷史上的活化石。

小　　龙：那现在还有人在用雕版印刷吗？

张教授：有啊。虽然雕版印刷不再大规模应用，但小范围还在使用，比如民间印家谱，过年印传统年画。

大　卫：要是能亲眼看看雕版印刷就好了。

张教授：当然能了。你们可以去扬州的中国雕版印刷博物馆，在那里可以了解雕版印刷的工艺流程。你们还可以去桃花坞木刻年画社学习年画制作。

小　龙：那太好了，谢谢张教授。

大　卫：张教授，桃花坞木刻年画社就在苏州吧？

张教授：没错，桃花坞年画就是我们苏州的民间艺术，现在还有桃花坞大街呢。当年桃花坞年画最昌盛的时候，画坊就有百余家。桃花坞的套色年画是中国南方年画的代表。现在年画已经成为非物质文化遗产了。有不少年轻人喜欢去年画社体验学习，你们也可以去看看。

小　龙：好的，我们一定找时间去体验学习。

注释：

① 四书五经：中国儒家经典书籍。四书指的是《论语》《孟子》《大学》和《中庸》，五经是《诗经》《尚书》《礼记》《周易》和《春秋》。

② 年画：中国特有的一种民间绘画体裁，也是常见的民间工艺品。年画大都新年时张贴，装饰环境，含有祝福新年吉祥喜庆之意。
③ 信笺：信纸的古代说法。

Engraved Block Printing

Xiaolong: Prof. Zhang, you mentioned that the technique of engraved block printing helped to promote public literacy and culture diffusion. Is it a very complicated technique?

Prof. Zhang: Yes, it is. The process of printing is time-consuming and labour-intensive. Nevertheless, you can produce books on a large scale through this method.

David: How many steps are involved in this technique?

Prof. Zhang: Quite a few. Step 1: transcribe a manuscript onto the paper; Step 2: paste the reverse transcription onto a wooden block; Step 3: engrave the characters or pictures on the block; Step 4: brush ink on the raised characters and then place a piece of paper on the block to get a printed sheet. Finally, bind the printed pieces into a book.

David: Well, it does sound complicated.

Xiaolong: Of course. But it's far more efficient compared to the practice of transcribing books manually, as many copies can be printed in a short time. I reckon it's more productive and effective to spread knowledge.

Prof. Zhang: You're right. It's a remarkable technique.

Xiaolong: It seems that the printed works in the Song Dynasty are especially famous. Why is that?

Prof. Zhang: This is mainly because the economy and technology of the Song Dynasty were well developed despite its political weakness and a small territory. The literature and art also greatly flourished, and thus books were in great demand.

David: So that's why books printed then are quite famous, right?

Prof. Zhang: Yes. Another factor also contributed to their fame. Before the Song Dynasty, there were no standardised typesetting and engraving processes. However, supervised by the Imperial Academy, the whole process and methods were standardised in the Song Dynasty, further improving the printing techniques. It was this standardisation of printing technique that helped print high-quality books.

Xiaolong: So, the works then are the best examples to study engraved block printing. It must have taken a long time for the craft to be well-developed, right?

Prof. Zhang: Yes, it had experienced the stages of seals, stone rubbings, and wood rubbings before it was well-established.

Xiaolong: Prof. Zhang, when did seals appear?

Prof. Zhang: They first appeared more than 2,700 years ago in the Eastern Zhou Period according to the unearthed relics and written records. Seals were initially used as evidence for doing business and later became a symbol of authority.

Xiaolong: They're still in use now. I know the seal is a unique way of Chinese signature.

David: Oh, it's different from our signature. We use autographs instead.

Prof. Zhang: Yes, it's different from the way of signature in your country. In China, seals have greatly influenced calligraphy. Enlightened by the seals, people started to engrave some famous calligraphy works on stone tablets using the intaglio engraving technique more than 2,000 years ago. In this way, calligraphy learners could copy famous calligraphic works. Later on, some whole books were engraved on the tablets.

Xiaolong: I guess people can get rubbing copies from the stone inscriptions.

Prof. Zhang: Yes. Stone rubbing first appeared during the Eastern Jin Dynasty, 1,600 years ago. It allowed more people to appreciate and learn calligraphy from tablets. You could lay a piece of paper on a stone tablet, and then tap it with ink sachets evenly to get a piece of rubbing.

Xiaolong: I know it. Most of the model calligraphy today are actually the rubbings. What about wood rubbing? Does it mean to engrave characters on a wooden block?

Prof. Zhang: Yes. It's similar to engraved block printing. You may lay a piece of paper on the wooden block, and then tap it with ink sachets to transfer the engraved words or pictures onto the paper.

David: It should be much easier to engrave characters on the wooden board.

Prof. Zhang: Absolutely. Another technique is stamping. It also had some connection with engraved block printing. It was often used to print images of Buddha for Buddhists, with engraved characters and patterns on the wooden blocks. The difference is that you have to press the printing block into the ink paste first before pressing it down onto the paper.

David: It feels like stamping a Buddha image.

Prof. Zhang: You're right. Precisely.

Xiaolong: Then it developed into engraved block printing, right?

Prof. Zhang: Yes. It came after the stone and wood rubbings. Then, a more convenient technique, clay or

	wood movable-type printing appeared. By using movable-type characters, it can save time in the process of block making.
Xiaolong:	It emerged in the Song Dynasty, right?
Prof. Zhang:	Right. The business and technology were booming at that time. Banknotes, the compass, wind-driven waterwheels, and spinners were widely used according to the records of the scholars then. The related knowledge could be passed on thanks to the large-scale printing.
Xiaolong:	However, the movable-type printing didn't become very popular after its invention.
Prof. Zhang:	There are some reasons for it. The first is about the technique itself. You know, factors such as different time and condition may cause certain changes in the clay and wood. If the characters are made separately, even tiny expansion and contraction during the printing process may result in uneven height and size, making it hard to get high-quality works.

Xiaolong: I see. The limitations of material made it unpopular.

Prof. Zhang: It's one of the reasons. Secondly, though there were many types of printed products, the contents updated slowly in ancient times, and the categories of printed books were relatively fixed. They included Confucian classics like the Four Books and the Five Classics[1] and other classics of Buddhism and Taoism. Moreover, the printed works for daily use such as New Year pictures[2], letter heads[3], account books, and calendars did not change much, so it was not necessary to change the engraved blocks.

David: Ah, I see. The relatively unchanged contents made it less demanding, so the advantages of movable-type printing didn't stand out.

Prof. Zhang: Right. Another important reason was that most printing craftsmen were uneducated, so it was difficult for them to do the typesetting.

Xiaolong: You mean there was a lack of technical drivers and social need for its large-scale development, right?

Prof. Zhang: Yes. Besides, after its invention, the engraved block printing was introduced to Korea and Japan, as well as other areas in Asia and Europe along the Silk Road. It was acclaimed as a living fossil in the field of printing for its long history and great influence.

Xiaolong: Is it still in use now?

Prof. Zhang: Of course. It's no longer applied massively, but still in use on a small scale such as printing genealogical records and traditional Chinese New Year pictures.

David: It would be great if we could see the actual process of the technique.

Prof. Zhang: Well, you can go to China Engraved Block Printing Museum in Yangzhou and learn more about the printing procedures there. And you can also visit the Taohuawu Woodcut New Year Picture Society to learn how to make New Year pictures.

Xiaolong: That's great. We'll go there. Thank you, Prof. Zhang.

David: Prof. Zhang, the society is in Suzhou, isn't it?

Prof. Zhang: Right. The New Year picture is a folk art in Suzhou. There's still a street called Taohuawu. Back in the heyday, there were over 100 workshops there. Their colourful New Year pictures were quite representative in southern China. Now, this tradition has become an intangible cultural heritage. Young people like to try their hands at the craft in the society. You should give it a try.

Xiaolong: Yes, we'd love to.

Notes:

1. The Four Books and the Five Classics: Chinese Confucian classics. The Four Books refer to *The Analects of Confucius, Mencius, The Great Learning,* and *The Doctrine of the Mean.* The Five Classics refer to *The Book of Poetry, The Book of History, The Book of Rites, The Book of Changes,* and *The Spring and Autumn Annals.*

2. New Year pictures: They are a special category of folk paintings in China. They are mostly used as decorations during the Chinese New Year and other festivals, for good luck and blessings.

3. Letter heads: It's an old-school term for letter paper in China.

制版

> 小龙和大卫来到扬州中国雕版印刷博物馆,向李馆长了解雕版印刷工艺流程。

小　　龙:您好,李馆长。我是小龙。这是我的好朋友大卫。

大　　卫:李馆长,您好!

李馆长:欢迎欢迎!张教授说你们想了解雕版印刷术。

小　　龙:是的,我们想看看雕版印刷的实物,也想请教一些具体问题。

李馆长:那我们一边参观,一边了解吧。我们这里还有一个雕版印刷体验馆,你们可以亲自做一回雕版印刷匠。

大　　卫:那太好了。

李馆长:我们先去看看雕版印刷的版片吧,这是最重要的。

> 他们来到雕版印刷展厅,大厅入口处能看到一个雕刻的反写大字"雕"。

大　　卫：这是什么字啊？

小　　龙：是反写的"雕"字吗？

李馆长：是的,雕版印刷需要阳文反刻。这样印出来的字才是正常的样子。等会儿你们去体验就清楚了。

> 三人一起走进版片库。一排排版架上整整齐齐摆满了版片,每个版片都比常规的书大一点,五六厘米厚。

小　　龙：哇,这么多！这里有多少版片呀？是不是都很古老？

李馆长：有20多万片。最早的是明代的,有700多年历史。最晚的是清代的,至少也有200年了。

大　　卫：这些版片现在还能用吗？

李馆长：这些原件都是珍宝,要保护好。现在一般用复制版片。现在这里是仓储式陈列,这种封闭性展示可保持版片室恒温恒湿,有利于长久保存。

大　　卫：收集到这么多版片应该很不容易吧。

李馆长：你说得很对。历史上,江南的刻书中心集中在南京、苏州和扬州。最兴盛的时候，扬州有上千家刻家。但随着现代印刷业的兴起，雕版印刷很快被替代，版片有的销毁，有的散失了。

小　龙：多可惜呀。

李馆长：但值得我们扬州人骄傲的是，我们将这一古老技术全面保存了下来。1960年，扬州市集中雕版印刷艺人，修复古版，并进行古书印刷。后来，江苏省内搜集整理的20多万片古旧版片也被集中到扬州统一管理使用。这其中包括丛书57种，单行本125种，

制版　Block Making

共 8900 多卷。2005 年，我们博物馆成立，这些古旧版片就转移到这里保管了。

小　龙：这些版片是怎么制作的？

李馆长：说到制版，得先从怎么挑选版材说起。选材的标准是硬度适中、纹理细滑。最常用的版材有梨木和枣木。

大　卫：制版过程很复杂吗？

李馆长：工序并不复杂，但选什么材质的树，什么时候取材特别重要。比如说，我们最喜欢用的梨木，上墨后固色特别好，不会晕染。梨木需要在夏天选材，腊月伐树。夏天我们要去找树皮发黑、大小合适的树。等到腊月树收浆的时候再伐，这样的梨木最适合做版片。

大　卫：那制版需要几道工序呢？

李馆长：一共有四道工序，第一道工序叫锯解。锯的时候要顺着纹理来，这样木板不容易开裂。第二道工序是浸沤，然后是干燥处理和平板两道工序。

小　龙：浸沤就是放到水里泡吗？

李馆长：是的，这是不可缺少的一道工序。

大　卫：浸泡后不是更潮湿了吗？为什么不直接晒干呢？

李馆长：木料经过浸沤后再进行干燥处理，不易变形。浸沤

　　　　　　还能将木板中的糖分泡出来，这样可以防虫。

小　　龙：看来每一道工序都很必要。干燥就是在阳光下晾晒吧？

李馆长：那样可不行，要自然干燥。把浸沤后的木板放在无直射光的通风处，每层之间用长木条垫平，这样可以使其干燥均匀。

大　　卫：那最后一道工序是平板了？

李馆长：是的。平板是将木板两面刨平整，然后用刮刀顺着纹理仔细刮平。制版的四道工序都非常重要，处理好就能确保做出高质量的版片。好的版片能使用上千次，而且吸墨均匀，字迹清晰。

小　　龙：版材还会影响墨的使用，这可没想到。

李馆长：墨是雕版印刷的主要媒介之一。有了墨，版上的图文才能转印到纸上。走，去看看我们馆收藏的墨。

Block Making

> Xiaolong and David arrived at China Engraved Block Printing Museum in Yangzhou and met with the curator, Mr. Li to learn more about the technique.

Xiaolong: Good morning, Mr. Li. I'm Xiaolong, and this is my friend David.

David: Nice to meet you, Mr. Li.

Mr. Li: Welcome! Prof. Zhang said you'd like to learn about engraved block printing, right?

Xiaolong: Yes. We'd like to see some engraved block-printed products and we're curious about the technique.

Mr. Li: Okay. Let's talk about them while we are looking around. We also have an interactive experience centre, and you can do the printing yourselves there.

David: That's great! Thanks a lot.

Mr. Li: Let's see the blocks first. It's the most important step in engraved block printing.

> They spotted a large Chinese character 雕 (*diao*) raised and engraved in reverse on the wall near the entrance.

David: What's this character?

Xiaolong: Is the character written reversely? Why is it so?

Mr. Li: Well, characters in engraved block printing should be carved in reverse so that the print comes out correctly. You'll understand it when you try printing yourselves.

> Upon entering the block library, they saw countless blocks packed tightly on the shelves. The size of each is slightly larger than that of an ordinary book, about 5 or 6 centimetres thick.

Xiaolong: Wow, so many blocks! How many are there altogether? Are they very old?

Mr. Li: Yes. There are more than 200,000 pieces. The earliest ones were made in the Ming Dynasty 700 years ago.

Even the latest are at least 200 years old from the late Qing Dynasty.

David: Can they still be used to print books now?

Mr. Li: Yes, of course. But they're very precious and we'd rather preserve them than use them. In fact, we usually use copies of these blocks for printing. Here is an enclosed space for the storage and display of the blocks. We keep the room at a constant temperature and humidity.

David: It must be very hard to have collected so many of them.

Mr. Li: You're right. In the past, Nanjing, Suzhou, and Yangzhou were the three centres of engraved block printing along the Yangtze River. So, there were thousands of printing workshops in Yangzhou in its full bloom. However, in the last century, it was gradually replaced by modern printing, and some of the old blocks were destroyed or lost.

Xiaolong: What a pity!

Mr. Li: Indeed. However, we're very proud that Yangzhou is the only city to have fully preserved the technique. In 1960, the craftsmen were assembled here to restore the ancient blocks and print new books with them. Eventually, more than 200,000 blocks had been sorted out from different parts of Jiangsu Province. In total, there were blocks for 57 series and 125 singles, making over 8,900 volumes of ancient books. When the museum was founded in 2005, they were all transferred here for better and safer protection.

Xiaolong: How were they made?

Mr. Li: As for the block-making process, the first important thing is to choose the wood, which needs to be of moderate hardness and have a smooth texture. The commonly used wood is either pear tree or jujube tree.

David: Is it very hard and complicated to make blocks?

Mr. Li: Not really. But it's particularly important to choose the right trees and cut them at the right time. For example, our favourite pear wood can absorb ink very well without any blurring. Summer is a good time to pick the trees with black barks and of proper size. Then, the last lunar month in a year is the right season to cut them down when they're in their best condition for making blocks for printing.

David: How many steps are there to make blocks?

Mr. Li: Altogether four steps. The first step is to saw the wood with the grain, so that the blocks won't crack apart. The next step is to soak them in water, and the last two steps are to dry and plane them.

Xiaolong: Does soaking mean immersing wood blocks in water?

Mr. Li: Yes, it can't be skipped.

David: Aren't they wetter after soaking? Why not dry them in the sun directly?

Mr. Li: Because soaking can protect the blocks from deforming when dried. It can also remove the sugar out of the blocks, and thereby preventing worm infestation.

Xiaolong: Well, it makes sense. Are they dried in the sun?

Mr. Li: That won't work. We should spread out the blocks evenly in the place without direct light, and use long wood strips between each layer, letting them dry evenly.

David: So, the last step is to plane the blocks, isn't it?

Mr. Li: Yes. The two sides of them should be planed evenly and polished carefully with the grain with a scraper. It's an extremely important step for good blocks to absorb ink well with clear prints and can be used thousands of times.

Xiaolong: I didn't expect the blocks to affect the ink absorption.

Mr. Li: Well, ink is one of the main elements for block printing. It helps to transfer texts and pictures from the blocks to the paper. Let's have a look at the ink displayed in our museum.

松烟墨

> 李馆长带着小龙和大卫来到一间小小的展室,展柜里放着大小不一的墨块。墨块上有图案、文字,有的还描着金色。

大　卫:这些是墨吗?我觉得它们像是艺术品呀。

小　龙:是啊,没想到墨块这么漂亮。

松烟墨　Pine-Soot Ink

李馆长：我们这里收藏的是松烟墨。

大　卫：松烟墨？烟怎么能做成墨呢？

李馆长：能啊，不过制作工序挺复杂。这种墨的主要原料是松树燃烧后留下的灰，所以选好松树至关重要。第一步，要在松树根部凿个小洞，用灯慢慢烤，让松脂流净，这叫去胶。

小　龙：啊？不是应该先砍树吗？

李馆长：不能先砍树，先要在树上直接去胶。去胶很重要，如果处理不好，会影响墨的渗透性。第二步，用竹篾编成半圆形竹篷，内外都糊上纸，立在地面上，然后把竹篷一节节连结起来搭在工棚里。竹篷用土掩实以防漏烟，内部用砖砌出烟道。在竹篷上，每隔一段距离开个小孔，用于出烟。

大　卫：下一步呢？

李馆长：第三步是制烟。这时再将除去松脂后的松树砍倒，劈成小块，放在竹篷的一头用慢火烧上几天。烟会顺着烟道进入竹篷，附着在纸上。然后，停火几天，等竹篷冷却后进篷收烟。

小　龙：怎么收烟呢？

李馆长：用鹅毛制成的扫烟工具，将竹篷上的松烟扫进容器内。烟按品质可以分为三类，最好的靠近竹篷尾部，这种烟最细，用于制造优质的墨。竹篷中间一段较细，用于制造一般的墨。靠近燃烧处的最粗，用来制作印刷用墨。

大　卫：原来松烟墨不只是用来印刷呀。

李馆长：松烟墨是墨中佳品，用途很广。写字、画画、印刷都可以。

小　龙：印刷用的墨和平时写字用的墨有什么不同呢？

李馆长：两种墨的制法不一样。一个是制成墨块，一个是制成膏状。印刷用墨是膏状墨，只需要研细，然后加胶料和酒制成膏状，在缸内存放三冬四夏，使臭味完全散去。这种墨存放越久，墨质越好。墨使用时，加水充分混合后，用一种特制的筛子过滤，然后用来印刷。

大　卫：那写字、画画是用墨块吧？墨块怎么制作呢？

李馆长：是的。制作墨块，得将炼好的烟用细绢筛到缸中，倒入上等皮胶和麝香、冰片、丁香等中草药和香料搅拌均匀。这是制作过程中最神秘的环节，配方从不对外公布。

小　龙：看来是商业秘密了。

李馆长：没错。接下来是锤墨。将配好的墨粉在铁臼里捣研。之后加水揉搓、捶打，直至初步成型。然后是放入墨模，把墨锭表面压制上漂亮的图案，再摊开晾干。一两的墨块需要晾6个月。墨块越大，晾干需要的时间越长。

大　卫：看来制墨还真需要耐心呢。

李馆长：这还没结束呢。最后一步是修墨添金，用工具将墨块的毛边打磨、修平，除掉瑕疵。

小　龙：添金是用金粉吗？

李馆长：对，添金又叫描金。墨块出厂前，工人用金粉等颜料对上面的图案和文字进行描画、填彩。

大　卫：没想到制墨也这么复杂。

李馆长：松烟墨难制，造纸也一样不容易。我们一起去看看优质的连史纸吧。

Pine-Soot Ink

> Xiaolong, David and Mr. Li went to a small display room. In the showcases were various ink sticks with different patterns or Chinese characters, some even painted in gold.

David: Are they made of ink sticks? I think they're art pieces.

Xiaolong: Wow, I didn't expect ink sticks could be so exquisite.

Mr. Li: They're made of pine-soot ink.

David: Pine-soot ink? Do you mean soot can be made into ink sticks?

Mr. Li: Yes, it can, but the whole process is very complicated. Its main raw material is the pine ash after the pine wood is burnt, so it's important to pick the right tree. When a tree is chosen, the first step is to degum. That's to say, workers need to cut a small hole in the root of a pine tree and place a lamp into it to heat the tree. In this way, the pine resin could be slowly removed.

Xiaolong: Ah, don't they cut the tree first?

Mr. Li: No. Degumming should be done on a live tree, which is vital for the permeability of the ink. The second step is to build a smoking shed. They weave some bamboo strips first into several semi-circular canopies with pieces of paper pasted both inside and outside. Then they put them on the ground and connect them one after another, and cover soil at their bottom to prevent soot leakage. Finally, a long flue is built with bricks inside the canopies, and holes are made at intervals to vent the smoke.

David: What's next then?

Mr. Li: To make the soot. The degummed pine trees are chopped into small pieces and burned slowly with low fire at one end of the flue for a few days. The smoke will float out of the flue and stick firmly to the paper. After that, the fire is stopped for a few days to cool down before the soot is collected.

Xiaolong: How do they collect it?

Mr. Li: They use a special tool made of goose feathers to sweep the soot into a container. The soot fits into three categories in terms of its quality. The finest one, collected from the far end of the canopies, is the best for making premium ink. The moderately fine soot from the middle part is used to make ordinary ink, while the coarsest one near the fire is for printing ink.

David: Ah, I see. Pine-soot ink is not just for printing.

Mr. Li: No, it's not only for that. It's of excellent quality, widely used in writing, painting, and printing.

Xiaolong: What's the difference between the ink for printing and writing?

Mr. Li: They're made in different ways. One is the ink stick for writing and painting, and the other is the ink paste for printing. To make the paste, you have to grind the soot finely before mixing it with glue and rice wine. The newly-made paste needs to be deodorised in a tank for about three years. The longer it's stored, the better quality it has. Before printing, you have to dissolve it in water thoroughly and refine it with a specially-made filter.

David: And ink sticks are used for writing and painting, right?

Mr. Li: Yes. To make ink sticks, you have to use fine silk to filter the soot into a tank, and then mix it with ox-hide glue and some Chinese herbal medicines and spices such as musk, borneol, and clove. This is the most mysterious part of making ink sticks, and the details of the formula have always been kept secret.

Xiaolong: I see. It's a business secret.

Mr. Li: Sure. Then, the third step is to shape the ink. The workers need to grind the prepared ink powder in an iron mortar. Then, they press and stretch it until it's initially shaped. After that, they put the ink into a mould to imprint a beautiful pattern onto the surface and finally dry it. It usually takes 6 months to dry a stick of 50 grams. The larger it is, the longer it'll take.

David: Well, it does need patience.

Mr. Li: It's not finished yet. The last step is to trim the stick and colour it golden.

Xiaolong: Do you mean using real gold powder?

Mr. Li: Yes, the patterns and characters are painted with pigments like gold powder before they're sold.

David: I didn't expect that making ink is so complicated.

Mr. Li: It's difficult to make the pine-soot ink, and making paper is just as challenging. Now let's take a look at the high-quality *lianshi* paper.

连史纸

> 小龙、大卫跟着李馆长来到专门收藏手工纸的房间,看到架子上整整齐齐摆放着一叠叠纸,有的洁白如玉,有的微微泛黄。

李馆长:我问你们个问题,知道为什么现在很多纸质文献书籍得小心翼翼保管吗?

小　龙:时间长了,纸会变脆,就没法翻阅了。

李馆长:没错。这主要因为原材料和制造方法不同。现代造纸的木浆耐久性差,而且造纸中使用化学药剂会让纸酸化变脆。这是影响纸质文献保存的世界性问题。

大　卫:这么严重吗?纸不是可以保存很久吗?

小　龙:对呀,不是说"纸寿千年绢八百"吗?

李馆长:这话没错,但这里说的是中国传统手工造纸,没有添加任何化学药品。

大　卫:明白了。您是说雕版印刷用的纸?

李馆长：是啊，要印出质量上乘的书，或者是想要印出代代相传的书，就需要用好纸。

小　龙：原来是纸好宋刻本才流传到现在。李馆长，传统雕版印刷用的是什么纸？

李馆长：中国传统纸按原料可分为麻纸、皮纸、藤纸、竹纸。雕版印刷主要用的是竹纸。你们看，这里存放的都是竹纸。

大　卫：可是它们颜色不一样呢。有的白一些，有的黄一些。

李馆长：颜色偏黄的是普通竹纸，这种很白的纸就是我要介绍的连史纸。摸摸看，感觉怎么样？

小　龙：白的很柔软，黄一点儿的好像手感差一些。

大　卫：我来摸摸。嗯，确实不太一样。

李馆长：这白的就是著名的连史纸。这种纸历史悠久，产自武夷山①地区。连史纸细腻洁白，防虫耐热，有着"百年不褪色，千年不变黄"的美誉。

小　龙：那是怎么做到的呢？

李馆长：秘密之一就是竹子本身。工人们需要在农历正月砍伐嫩竹，此时竹浆很丰富。

大　卫：还有什么工序呢？

李馆长：选材只是第一步。制作工艺特殊，是连史纸品质优良的另一个原因。古时制造连史纸需要72道工序，出成品需要一年左右的时间。

小　龙：太不容易了。看来必须有足够的时间和耐心才能出好东西。

李馆长：是啊，在制备过程中，要先用石灰浸泡和过碱。这样做可软化竹丝，同时也去除糖分和淀粉，只留下竹纤维。

大　卫：看来要把事情做好不能着急。急了制不了好墨，急

连史纸　*Lianshi* Paper

了也造不出好纸。

李馆长：大卫说得很对。造纸过程分三个阶段：砍竹、作料、抄纸。采伐后的嫩竹要放入池塘浸泡，然后剥成竹丝再反复浸泡。最后洗净、去除杂质、晒干，成为"竹麻丝"。

小　龙：作料就是把它们变成纸浆，对吗？

李馆长：对的。这个过程最费时费工，每个环节都关系到是否能造出质量上乘的纸。竹料在发酵的石灰水中蒸煮后做成"丝饼"，放在山坡上。这些丝饼分两次进行长达半年的自然漂白，慢慢氧化成白色。

大　卫：半年真的很久呀。那下一步做什么呢？

李馆长：自然漂白后要把它们舂成细泥状，这个工序叫榨纸。纸榨完后放入料槽，用脚踩开，为最后的抄纸做好准备。抄纸前还需要一道重要工序——加纸药。

大　卫：古代造纸也要加药？

李馆长：对，纸药是天然的抄纸助剂，从一种植物黏液中提取。它可以让纸浆纤维在水中分散均匀，漂浮在水上。

小　龙：下面就该抄纸了吧？

李馆长：是的，抄纸是最考验技术、最需要经验的环节。过去只有技术熟练的老师傅才有资格抄纸。

大　卫：具体怎么做呢？

李馆长：用一种细竹丝编成的竹帘在纸浆池中轻轻荡抄，帘子滤水后会留下一层薄的纸浆膜。然后，再把帘子反扣过来，让湿纸落到木板上。等到湿纸叠积500张，再用木质杠杆式压榨机挤出大部分水分。

小　龙：那纸不会粘在一起吗？

李馆长：加过纸药就不会粘在一起了。烘纸师傅会用细竹签将纸逐张挑开，然后用刷子把纸贴到专门的烘纸土墙上。那是一种用土砖砌成的夹墙，在墙中生火。

大　卫：为什么不自然晾干呢？

李馆长：一是时间来不及，二是烘纸土墙很平整，可以把纸直接熨平。纸干后揭起，再剪裁整理。纸造好后需要存放一段时间才能使用。

大　卫：那是为什么？

李馆长：我们称作去火，主要是让纸舒展开，这样吸墨效果好。

小　龙：真没想到，生产一张纸需要那么长时间，工艺还那么复杂。这么好的古代技艺得好好保护和传承呢。

李馆长：是啊，不能有了现代技术就丢了祖先创造的传统技艺。2006年，连史纸制造工艺入选了首批《国家级非物质文化遗产名录》。

注释：
① 武夷山：位于江西与福建西北部两省交界处，是中国著名的风景旅游区和避暑胜地，也是世界文化与自然双重遗产地。

Lianshi Paper

> Xiaolong, David and Mr. Li went to the room where hand-made paper was stored and piles of paper were neatly placed on the shelves. Some were white and others yellowish.

Mr. Li: I have a question for you. Why do we take great care of paper documents and books today?

Xiaolong: Paper is very fragile and easily becomes brittle over time. Then the old book pages will be prone to tearing and difficult to turn when reading.

Mr. Li: You're right. The materials and methods used in modern paper-making are quite different. The wood pulp used today is of poor durability, and the chemicals tend to make the paper fragile. This poses difficulties for long-term storage of paper books and documents worldwide.

David: Is the issue so serious? I thought paper could last for a long time.

Xiaolong: Me too. I remember we have an old saying. It says that paper can last a thousand years and silk eight hundred, right?

Mr. Li: It's true. But this saying actually describes the quality of traditional Chinese hand-made paper, which contains no chemicals.

David: I see. You mean the paper used for the engraved block printing.

Mr. Li: Yes. With high-quality hand-made paper, we can print books and pass them on from generation to generation.

Xiaolong: So, that's why we can still read the books printed in the Song Dynasty. Mr. Li, what types of paper are used for engraved block printing?

Mr. Li: There are flax and hemp paper, bark paper, rattan paper, and bamboo paper according to the raw materials. Here is the bamboo paper, mainly used for engraved block printing.

David: But the paper here has different colours. Some piles are white and others yellowish.

Mr. Li: Well, the yellowish paper is ordinary bamboo paper, while the white paper is *lianshi* paper. Can you feel the difference of the two piles?

Xiaolong: Yes, the white paper feels soft while the yellowish one is quite coarse.

David: Let me see. They do feel different.

Mr. Li: The white paper is the famous *lianshi* paper. It has a long history and comes from Mount Wuyi[1]. *Lianshi* paper is fine and white. What's more special is that it can repel insects and withstand heat. It enjoys a good reputation for remaining unfaded after hundreds of years and for not turning yellow even in a thousand years.

Xiaolong: How exactly is it made to do that?

Mr. Li: One secret is the bamboo itself. It's usually cut down in the first lunar month when the tender bamboo has no leaves but plenty of pulp.

David: And what are the other secrets?

Mr. Li: Another is the sophisticated craftsmanship. There were 72 steps to make *lianshi* paper in ancient times, and it often took a year for the final product.

Xiaolong: It's not easy. It seems only time and patience can make the best.

Mr. Li: Sure. The bamboo should be soaked in lime and treated with the alkali water. This helps soften the fibre and remove sugar and starch. That's why the paper won't easily break and can remain high quality over time.

David: It really takes time to do things well. Neither good ink nor high-quality paper can be made in a hurry.

Mr. Li: You're right. There are three major steps in paper-making: bamboo-cutting, pulp-making, and pulp-film-filtering. In the first step, the newly-cut bamboos should be soaked in a pond until they're totally rotten with only fibres left. Then, the workers should wash them to remove impurities and let them dry in the shape of bamboo thread.

Xiaolong: So, pulp-making means smashing the thread into the pulp?

Mr. Li: Yes, it's a time-consuming and labour-demanding job. Each step is crucial for making high-quality paper. As mentioned before, it should be boiled in fermented limewater. Then, it needs to be placed on the hillside to oxidise naturally. This step has to be done twice and it takes a total of half a year. Gradually, it'll turn white in the air.

David: Wow, it's a long time. What happens next?

Mr. Li: They're then mashed into fine paper mud and placed into the pulp pool. Finally, another important ingredient, a paper catalyst, is added to the pool before filtering.

David: What's a paper catalyst?

Mr. Li: It's a natural substance extracted from the mucus of a certain plant. It helps disperse the pulp-film evenly on the water surface.

Xiaolong: Now it's time for pulp-film filtering, right?

Mr. Li: Yes, it's key to paper making and requires much skill and experience. Only the most experienced workers can do it.

David: How to filter it then?

Mr. Li: The workers would use a bamboo filter made of fine bamboo slivers. They gently swing it in the pool, leaving a thin layer of pulp film on the filter. Then, turn it over to invert the wet paper onto a wooden board. When there are 500 pieces in a pile, a special wooden lever press is used to squeeze the water out as much as possible.

Xiaolong: Don't they stick together?

Mr. Li: No, they don't, thanks to the catalyst. The paper baker would use a thin bamboo stick to pick up each sheet and place it on a baking wall. It's a kind of clay-brick hollow wall with fire inside.

David: Why not dry them naturally?

Mr. Li: It takes too long time. And more importantly, the wall can dry the paper evenly. After it's dried, the paper can be cut into the same size. Moreover, it should be kept for some time before being used.

David: Why?

Mr. Li: We call it paper-cooling. Then the paper can stretch out itself for better ink absorption after that.

Xiaolong: I really didn't expect the process of paper making to take such a long time and involve so many complicated procedures. This papermaking technique should be preserved and passed on.

Mr. Li: Yes, you're right. The traditional technique created by our ancestors can't be lost, even with advanced modern technologies. In 2006, this technique of making *lianshi* paper was selected as part of the first batch of China's National Intangible Cultural Heritage.

Note:

1. **Mount Wuyi**: Located at the border of Jiangxi Province and the northwest Fujian Province, it's a famous scenic spot and summer resort in China, and also a UNESCO's dual heritage of world culture and nature.

写版与校正

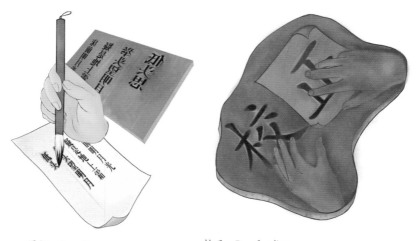

写版 Sampling　　　　　　校正 Proofreading

李馆长带着小龙和大卫来到雕版印刷体验馆。体验馆入口处放着一些大小不一的木板和宣纸,宣纸上写有毛笔字。

李馆长:我们先参观体验一下雕版印刷的主要环节,然后再去看雕版印刷作品,那样印象会更深刻。先从写版开始吧。

小　　龙：李馆长，雕版印刷的写版是写在纸上还是板上呢？

李馆长：你们认为写在纸上容易，还是板上容易？

小　　龙：纸上写好像更容易一些。

李馆长：是的。不过，雕版印刷的写版两种方法都用。

大　　卫：用毛笔抄写整齐也不容易吧？把字写歪了怎么办呢？

李馆长：在写版用的纸上用虚线打好格子，就可以避免这个问题。我们再说说要写的字吧。你们看，这里有几个"永"字，你们知道都是什么字体吗？

小　　龙：这个应该是小篆吧。印章上最常见。

李馆长：对的。下一个呢？

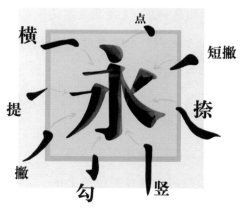

"永"字宋体　Song Typeface of Character "永"

大　卫：这应该是楷书。我练过。

李馆长：正确。最后一个呢？

小　龙：这个我熟悉，是最常用的宋体。

李馆长：是的。现在出版物一般都是用宋体印刷。宋体的特点是横细竖粗、末端有装饰。点、撇、捺、钩等笔画都有尖端。

大　卫：宋体是宋朝人发明的吗？

李馆长：对啊。我问你们一个问题，为什么宋朝人要发明宋体字？

大　卫：是为了好看吗？

李馆长：当然不是。

小　龙：那就是为了刻版好刻。

李馆长：对了。你们想一想，毛笔字是不是讲究起笔收笔啊？刻字就很难做到，也很费时。于是，就出现了利于刻字的宋体。

小　龙：我还以为宋朝人比较刻板，所以才出现了这么硬直的字体。

李馆长：不是的。其实，为了刻印方便，字体才逐渐从楷体演变成专门用于雕版刻印的宋体。应该说是雕版印刷促成了宋体字的成熟。宋朝的雕版印刷品质量上乘，所以后人很喜欢翻印宋刻本。

大　卫：哦，原来是这样。

> 李馆长、小龙和大卫来到一位写版师傅跟前。师傅手握毛笔，正在一张宣纸上誊写《静夜思》。

大　卫：这首是李白的诗，我学过，"床前明月光，疑是地上霜，举头望明月，低头思故乡"。

李馆长：大卫，你会背李白的诗，真不错。

大　卫：我喜欢中国诗。我现在学着背一些简单的诗，这对我理解中国文化很有帮助。

小　龙：李馆长，师傅们是在写版吗？师傅们的毛笔字可真漂亮呀。

李馆长：是啊。他们正在写《唐诗三百首》的版，自然不能少了李白的诗呀。你看，这叫正写，就是在纸上正常写字。

大　卫：正写？难道还有反写？

李馆长：说对了，是有反写。一般来说，为了保证全书字体一致，一本书只能由一个人抄写。

大　卫：一个人抄一整本书？要好长时间吧？

李馆长：是的。但抄完之后，用雕版印刷就可以印成千上万本。雕版印刷术对古籍保存贡献很大。

小　　龙：的确。李馆长，我有个问题。如果字写错了怎么办？需要重新抄一份吗？

李馆长：不用那么麻烦，我们有专门的校正师傅做校正。通常错字改过来就行。写样之后，誊写的内容需要由专人仔细校对，将错别字找出来并及时改正。

大　　卫：如果错误很多是不是就得重新写了？

李馆长：这个一般不会。写版师傅都技术熟练，错误率会控制在万分之一以内，不会出现很多错别字。

大　　卫：好厉害啊！

李馆长：如果出现错别字，校正师傅就用白纸把错字盖上，重新誊写正确就可以了。纠错讲究又快又准，得由有经验的师傅来做。

小　　龙：那下一步该做什么？

李馆长：你们自己来体验一下，就知道做什么了。

Sampling and Proofreading

> Xiaolong, David, and Mr. Li went to the experience centre. They found at the entrance some blocks of different sizes and some pieces of *xuan* paper with Chinese characters written in calligraphy.

Mr. Li: We'll go through the main procedures of engraved block printing here and then see the products of the printing. In this way, I believe you'll have a deeper impression. Now let's start with the sampling.

Xiaolong: Mr. Li, do we write characters on the paper or the block?

Mr. Li: Which one do you think is easier?

Xiaolong: Writing on the paper, of course.

Mr. Li: Right. Both are used in engraved block printing.

David: It's hard to transcribe correctly and exactly, isn't it? What can we do if the words are written crookedly?

Mr. Li: No worries. The thin paper for the transcription is usually gridded with dotted lines, so you can write them in the right place. Let's talk about the characters. Look, here are different fonts of the character *yong* (永) . Can you recognise them?

Xiaolong: This font must be the Small Seal. It's popular for seal engraving.

Mr. Li: Correct. What about this one?

David: The Regular Script. I have practiced it in my calligraphy class.

Mr. Li: Good. The last one?

Xiaolong: I know it. It's the Song Typeface, the font most commonly used today.

Mr. Li: Right. The Song Typeface is very popular in printing books, magazines, and newspapers. It features fine horizontal strokes and thick vertical strokes with decorative clear ends.

David: Was it invented in the Song Dynasty?

Mr. Li: Yes, it was. I have a question for you. Why was the typeface invented then?

David: To make the printing more beautiful?

Mr. Li: Of course not.

Xiaolong: To engrave easily?

Mr. Li: Absolutely right. Just think about the Chinese calligraphy. It requires beautiful ends in writing characters, doesn't it? But it's hard and time-consuming when engraving characters. The Song Typeface makes the engraving of characters easier. That's why the Song Typeface becomes popular.

Xiaolong: I thought this kind of blocky font was invented because of the traditional culture in the Song Dynasty.

Mr. Li: No. In fact, the font had gradually changed from the Regular Script into the Song Typeface, especially when used for the engraving. That is to say, it was the practice of engraving that helped develop this font. With upright characters, the books printed in the Song Dynasty are of high quality, and they're often reprinted by later generations.

David: Oh, I see.

> Xiaolong, David and Mr. Li went to a sampling worker, who was transcribing the poem "A Tranquil Night" on a piece of paper with a brush pen.

David: I've learnt this Tang poem. It's written by Li Bai. "Before my bed a pool of light. Can it be hoarfrost on the ground? Looking up, I find the moon bright. Bowing, in homesickness I'm drowned."[1]

Mr. Li: Wow, David, it's so amazing that you can recite Li Bai's poem.

David: I love ancient Chinese poetry. I'm trying to memorise some simple Chinese poems. They do help me understand Chinese culture.

Xiaolong: Mr. Li, are they doing the sampling? They have truly beautiful calligraphy.

Mr. Li: Yes. They're doing sampling work of *Three Hundred Poems of the Tang Dynasty*. Li Bai is the first poet we should include in the collection. Look, this is called positive writing. It refers to the normal writing on the paper.

David: Positive writing? Do you mean there is negative writing?

Mr. Li: Right. It's called mirror-image writing. Most of the time, one book is copied by the same scribe to keep the whole book in the same style.

David: Does it take a long time to transcribe a book?

Mr. Li: Yes, it takes much time. However, once the sample is completed, thousands of copies can be printed. Engraved block printing made a great contribution to the preservation of Chinese classics.

Xiaolong: Indeed, but Mr. Li, I'm curious. Do we need to rewrite the whole piece if we write the wrong words?

Mr. Li: No, not necessarily. You just need to make the corrections. We have professional proofreaders to help you. They'll proofread the transcription after the sampling, find out the errors, and correct them immediately.

David: If there are too many mistakes, does it mean the scribes have to redo the whole book?

Mr. Li: Don't worry. The scribes are very experienced. The error rate of their handwriting is usually controlled within one in ten thousand, so there are few errors.

David: That's incredible!

Mr. Li: The proofreader just needs to cover the wrong character with a patch of paper and rewrite it correctly when they find the wrong one. This job requires rich experience in proofreading since it needs to be done quickly and accurately.

Xiaolong: What's next then?

Mr. Li: Next, give it a try yourselves. You'll understand it better once you do it yourselves.

Note:

1. It was translated by the famous translator, Xu Yuanchong.

上样

李馆长拿来两块梨木板和两张誊写好的稿子。

李馆长：我给你们找了两块板子。来，一人一块。
大　卫：是把稿子贴在板子上吗？

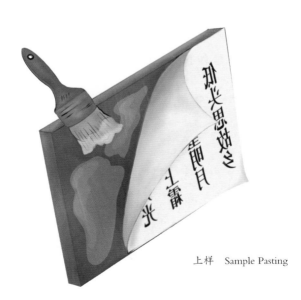

上样　Sample Pasting

李馆长：是的，这叫上样。

小　龙：李馆长，这个像酸奶的东西是什么？

李馆长：这是由面粉加水熬制的浆糊。我们小时候常用，相当于现在的胶水。

大　卫：是把它刷在板子上吗？

李馆长：是的。但要小心，要均匀地、薄薄地刷。

小　龙：您看，这样行吗？

李馆长：非常好。接下来，把这两首唐诗的稿子平放在木板上。注意，有文字或图案的一面朝下。

大　卫：为什么要反着放呢？

李馆长：还记得刚才我说的，雕版印刷技艺就是正字反刻，这样印刷出来的字才是正的。

大　卫：哦，想起来了。我贴好了，下一步该做什么呢？

> 李馆长递给他们一人一把刷子。

李馆长：现在，用刷子从中间向四周轻轻地刷，挤去纸与木板之间的气泡。

小　龙：这有点像给手机贴膜呀。

李馆长：是的。上样，目的就是让纸与木板紧密贴合在一起。

如果还有小气泡，要用针扎破，然后轻轻刷平。

大　　卫：那接下来应该是刻版了吧？

李馆长：现在还太早。你们看，纸还有点湿，等干了，用手轻搓纸面，让它变薄，然后刷一层油。

小　　龙：为什么还要刷油呢？

李馆长：刷过油之后，字体会更清晰，刻印时就会看得更清楚。

小　　龙：没想到上样还有门道呢。

李馆长：那当然了。上样是一道关键工序，上版的质量会直接影响到刊刻效果。

大　　卫：如果这一步没有做好，会有什么麻烦？

李馆长：会很麻烦。如果誊写的稿纸与木板黏合不牢固，刊刻时纸张浮起，会造成下刀不准。如果稿纸贴偏，刊印出来的字或图案也就歪斜了。要是套印画稿，上样要求就会更严格。不能有一丝一毫的差错。

小　　龙：看来所有工序都得认真仔细。

Sample Pasting

> Mr. Li handed over two pear wood blocks and two transcribed pieces of paper to David and Xiaolong.

Mr. Li: I got you two blocks. Here you are.

David: Do you mean that we'll paste the transcribed paper on it?

Mr. Li: Yes, the step is called sample pasting.

Xiaolong: Mr. Li, what's this yoghurt-like stuff?

Mr. Li: It's a sort of paste made from flour and water. It works just like glue.

David: So, we're going to paste it on the block?

Mr. Li: Exactly. Now brush a thin layer of paste on the block evenly.

Xiaolong: Look, Mr. Li, is this all right?

Mr. Li: Yes. And next, place the transcriptions on the block.

Make sure the text or pattern is facing the block.

David: Why do we do this?

Mr. Li: Remember what I mentioned? Only if the texts or patterns are placed facing the block will the print come out correctly oriented.

David: Ah, got it. I've finished. What's next?

> Mr. Li handed each of them a palm brush.

Mr. Li: Now, gently brush the paper from the middle to the sides. This helps get rid of the bubbles between the paper and the block.

Xiaolong: It looks like applying a screen protector on a mobile phone.

Mr. Li: Quite similar, indeed. Then you press the paper firmly onto the block. If there are still bubbles, you can use a needle to prick them and brush the paper evenly again.

David: Can I start to engrave now?

Mr. Li: No need to hurry. Look, the paper is still wet. When it dries, you can gently rub the paper to thin it out and then apply some oil.

Xiaolong: Why do I need to apply oil?

Mr. Li: The oil helps make the text or pattern clearer when you start engraving.

Xiaolong: I never thought sample pasting would have so many intricacies involved.

Mr. Li: It's essential for printing and really influences the quality of the final prints.

David: What if something goes wrong?

Mr. Li: Well, if it's not securely stuck to the block, the paper will shift as you engrave, throwing off the carving position. And if the sample is pasted crookedly, the printed work will turn out crooked as well. Furthermore, when it comes to the overprinting of woodcuts, getting the sample pasting right is even more critical, leaving zero room for errors.

Xiaolong: I see. Every step counts.

刻版

> 一位师傅手握刻刀，熟练地在上好样的木版上刻着。看着他面前放着一排大大小小形态各异的刀具，小龙和大卫非常好奇。

大　　卫：哇，要用到这么多刀呀。

李馆长：是的，每把刀都有专门用途。根据功能不同，雕刻刀可以分为刻刀、铲刀、削刀，还可以根据形状分为曲刀、平口刀。

小　　龙：李馆长，这把刀长得好奇怪呀。

李馆长：猜猜它叫什么刀？

小　　龙：握着很舒服，是握刀吗？

李馆长：形容得很贴切，不过它不叫握刀，叫拳刀。

大　　卫：拳刀？为什么这么叫呢？它长得也不像拳头啊。

李馆长：这是因为雕刻过程中，工人手握刀用力下刻时，手

刻版　Block Engraving

会收成拳头状。拳刀是雕版中最重要的刀具。

小　　龙：原来是这样。这位师傅刻得好熟练呀。但大部头的书刻起来需要好几年吧？

李馆长：是的，刻书需要耐心和恒心。

大　　卫：李馆长，我刚才在门口看到好多木板和写了字的宣纸。那是干什么用的？

李馆长：我们这个馆是雕版印刷体验馆。在这里，游客可以自行选板，请师傅制作自己喜欢的文字或图片，留作纪念。也可以在师傅指导下，自己动手刻印。

小　　龙：这太有趣了。我能试试吗？上中学的时候我学过刻图章呢。

李馆长：有刻字的基础，应该能行。你是刻字还是刻图呢？

小　　龙：想刻一个"福"字，我还想自己写"福"字，可以吗？

"福"字阳文反书　　Reversely Pasted "福"

李馆长：没问题。自己动手，更有意义。

大　卫：那我来上样。

李馆长：好呀，小龙写字雕刻，大卫上样，两人合作完成一个作品。

> 小龙用毛笔在纸上认真地写了一个"福"字。大卫把它反贴在木板上。等木板上的稿纸干透，大卫洒了一点儿水，然后轻轻把稿纸搓薄，最后用毛笔沾油把有字的地方再描了一遍，字迹变得非常清晰。

大　卫：你们看，木板上的"福"字很清楚了。

李馆长：不错，小龙字写得很棒，大卫上样做得也很好。

小　龙：现在可以刻字了吗？

李馆长：嗯，先用直尺比对着，用刀把有字的地方划出来，剔除空的部分，再用拳刀慢慢刻字。刻字分发刀和挑刀两个工序，要用两种不同的刀具。发刀工序一般是先刻字的左侧和下口，然后再掉个头，刻字的右侧和上口。

大　卫：发刀用的刀具真薄啊，很锋利吧？

李馆长：确实很薄，可以刻得很深，一般是由有经验的师傅发刀，保证刻得横平竖直，字迹清晰。

小　　龙：那什么是挑刀工序呢？

李馆长：挑刀工序一般是徒弟来完成，要由徒弟用厚一点儿的刀在空白处一行行下刀。

小　　龙：哦，明白了。大卫，我们一起来刻吧。

> 小龙仔细运刀，把"福"字一点点儿仔细刻了出来。李馆长请来一位刻版师傅帮忙，把字修整得更清楚整齐。

李馆长：刻得很不错嘛。现在用锯条把多余的部分锯掉，再用刮刀刮平。看看会不会更漂亮些？

小　　龙：还真是这样，现在看起来好多了。

李馆长：还没结束呢。板上有你手上的汗渍，凹槽里还有好多碎木屑，需要好好清洗一下。

小　　龙：好的，我去用水冲一下。

李馆长：不能用水冲。要用白毛巾盖到木板上，然后浇上开水闷三到五分钟，最后仔细擦干净就可以了。

大　　卫：这个我来做。

小　　龙：刻版真有意思。

李馆长：来看看你们合作的"福"字，有什么发现吗？

大　　卫：字是反的，还凸起来了。

李馆长：对，这就叫"阳文反刻"。

小　　龙：字凸起来就是阳文，那凹下去就是阴文了？张教授说过，石碑上的碑文大多是阴文。

李馆长：是的，阳文和阴文的概念是相对而言。"阳文反刻"是雕版印刷的关键。然后才是下一步，刷印书籍。

小　　龙：这下我懂了。大卫，我们把这块版带回学校，多印一些"福"字，送给老师和同学们。

大　　卫：这个主意好。给大家送福。

李馆长：那你们去学学怎样刷印吧。

Block Engraving

> An engraver held a knife in his hand, skilfully carving a sample-pasted block. Xiaolong and David were very curious when looking at the knives of different shapes and sizes in front of him.

David: Wow, so many knives.

Mr. Li: Aye, different knives are used for different jobs, you see. Engraving knives can be classified into carving knives, shoveling knives and paring knives. And according to their shapes, you've got curved ones and flat ones.

Xiaolong: Mr. Li, this knife looks very strange.

Mr. Li: Have a guess what it's called.

Xiaolong: It feels good to hold. Could it be called a "grip knife"?

Mr. Li: Not a bad shot. But it's actually called a "fist knife".

David: Fist knife? But it doesn't look like a fist at all. Why this name?

Mr. Li: It's so named because it has something to do with how the engravers hold it. When they're carving on the block, their fists sort of clench around it. It's the most important knife for engraving.

Xiaolong: Ah, I see now. Look at him. How skilful he is! It must take years to carve a multi-volume work, eh?

Mr. Li: Yes, patience and perseverance are essential.

David: Mr. Li, I've noticed a bunch of wooden blocks and *xuan* paper with words written on them near the entrance. What are they used for?

Mr. Li: Well, those items in this experience centre are meant for visitors to choose their favourite characters or images. And visitors can create their own printed pieces later with their preferred words and images. They can ask our staff members for help or try their hands at engraving and printing on their own.

Xiaolong: That sounds fascinating. Can I give it a shot? I've learned seal engraving when I was in middle school.

Mr. Li: Sure. Do you want to engrave a character or a picture?

Xiaolong: I'd like to engrave and print the character *fu* (福). Can I do it myself?

Mr. Li: Certainly, DIY will be more fun.

David: I'll handle the sample pasting.

Mr. Li: Great. It's going to be a perfect piece. Xiaolong can write and engrave the character and David can do the sample pasting.

> Xiaolong carefully wrote the character *fu* on the paper with a brush pen. David pasted it onto the block. After the paper had dried out, he sprinkled some water on it and gently rubbed the paper. Then, after being brushed with some oil, the character became very clear.

David: Look, now the character stands out clearly.

Mr. Li: Well done.

Xiaolong: Can I start the engraving then?

Mr. Li: Of course. First, use a knife and ruler to frame the character and cut its border. Then use a fist knife to carve it slowly. There are actually two steps to carve the character. The first step is called carving-in, *fadao*

(发刀) in Chinese. It means to carve the left side and the lower part of the character, and then turn around the block to carve its right side and the upper part.

David: This knife looks quite thin. It must be very sharp, right?

Mr. Li: Yes, this type of knife can cut deeply. The step of carving-in is usually done by an experienced worker to ensure that each character can be carved clearly.

Xiaolong: What's the second step?

Mr. Li: It's the step called carving-out, or *tiaodao* (挑刀) in Chinese. It's usually done by an apprentice. It means using a thicker knife to carve away the blank parts line by line.

Xiaolong: I see. David, let's do the block carving together.

> Xiaolong carved the character *fu* carefully. Mr. Li invited an engraver to help him make the character tidier and clearer.

Mr. Li: Good job. Now cut off the extra part with a saw, and smooth it out with a scraper. Now, look, it's more beautiful, isn't it?

Xiaolong: Wow, much better.

Mr. Li: It's not finished yet. There are some sweat stains on the board and splinters in the grooves. You need to clean it carefully.

Xiaolong: Okay, I'll rinse it with water.

Mr. Li: No, you shouldn't do that. Just cover it with a white towel and pour some hot water over it. After about five minutes, wipe it clean.

David: Ah, I see.

Xiaolong: This is very interesting.

Mr. Li: Did you notice anything different about the character?

David: Yes. It's reverse-engraved and raised.

Mr. Li: You're right. It's called a reverse engraving of a *yang* script.

Xiaolong: If a raised one is *yang* script, can we call a concave character *yin* script? Prof. Zhang said that most of the inscriptions on the stone tablets are *yin* scripts.

Mr. Li: Yes. The concepts of *yang* and *yin* are opposite. The reverse engraving of a *yang* script is key to block printing. Now, let's move to the next step: book printing.

Xiaolong: Got it! David, how about taking the block back to print more characters of *fu* and distribute them?

David: Good idea! Let's send more people our blessings.

Mr. Li: Now let's head to the printing zone.

刷印

小龙和大卫跟着李馆长来到刷印台前。

棕帚（左） 棕擦（右） Palm Brush (Left) Palm Smoother (Right)

小　龙：墨、纸、版片……咦，这两个工具是什么呀？

李馆长：这个是棕帚，用来给版片刷墨。另一个是棕擦，用来刷印纸。

小　龙：好像都是用棕榈叶做的。

李馆长：准确地说，是用山棕榈树的棕衣做的。

大　卫：看，纸的这边怎么固定在桌子上了？

李馆长：是这样的。刷印前要把纸和印版固定好。刷印时为了避免印版滑动，一般会用钉子固定印版，这叫"固版"。纸固定在桌子上，是为了压印翻页时方便，不打乱顺序。

小　龙：明白了。可是纸和刷印台之间怎么有个缝呢？

李馆长：刷印台的设计和布置是为了方便工作，提高效率。这个，一会儿你自己上台去印就明白啦。那我们现在就开始刷印吧。这里有唐诗的版片。

大　卫：我要印李白的诗。

小　龙：王维是我最喜欢的诗人，我想印王维的诗。

李馆长：可以。刊刻书通常要先刷红色，出红印校本，用于第一遍校对。如有错，校正后，出蓝印本，用于二校。二校完全没有问题了，才开始大批量墨印。今天我们用现成的版片直接墨印。

大　卫：我现在就开始刷墨了。

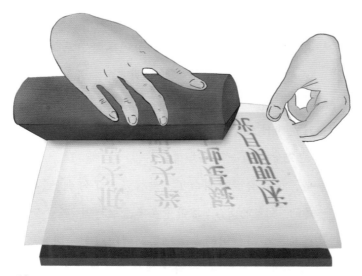

刷印 Brush Printing

李馆长：别着急，刷墨之前，先用清水将印版刷两遍。

小　龙：版片不是很干净吗？为什么还要刷清水呢？

李馆长：先用清水湿润版片，再正式刷色，更容易上色。

大　卫：哦，知道了。清水刷好了，现在可以刷墨了吗？

李馆长：可以了，但要讲究手法。你们看一下这个棕帚，能发现它有什么特点吗？

小　龙：它的头不是齐的，从里向外是阶梯状，好像有四层。

李馆长：是的。别看它很简单，但非常科学。这样刷墨会持续不断送墨，而且不会晕染，所以它也叫"均墨器"。

大　卫：真没想到，这个不起眼的刷子还很科学呢。

李馆长：是的。我们得佩服古人的聪明才智。开始刷墨之前，得先把棕帚放热水里泡一下。

大　卫：这是为什么呀？

李馆长：泡完你们摸一下就知道了。

大　卫：哦，很软呀。我明白了，这样是为了好刷墨。

李馆长：是的。刷墨的时候，棕帚先在墨盘中打圈，蘸少量墨汁。然后，还用打圈的方法把墨刷到印版上。同时还要注意，刷的速度要快，这样才能保证墨色的饱和度一致，印出来的东西颜色没有差别。

大　卫：那接下来，就要把纸放在上面了吧？

李馆长：对，这一步叫"覆纸"。像这样，左手捻拉住纸，平放到印版上，右手拿棕擦在纸背上刷。

小　龙：你们快看，开始有墨迹了。

李馆长：不要急着揭开。你看，这边墨色不够，得多擦几下。等印到纸上的字迹或图案完整清晰了，才可以揭纸。

> 大卫刷印的唐诗第一页已经清晰地印在纸上了，上面的墨迹是半湿的状态。

大　　卫：这怎么办呀？墨还没完全干呢。

李馆长：你们看，这条缝隙是不是该派上用场啦？

大　　卫：原来这条缝隙是让纸自然下垂晾干呀！这个设计真巧妙，很实用。

李馆长：是的，没错。但你只说对了一半，这不仅仅是为了晾干。我们现在这种印刷方法叫压印，纸是固定在桌子上的，印完了得有地方放纸呀。

大　　卫：哦，原来和这个工艺有关。

小　　龙：纸不固定在桌子上，就不需要这个缝隙了，对吧？

李馆长：对的。雕版印刷有两种印刷方式，一种是飞印，一种是现在这种压印。压印通常是用于套色印刷，这样固定纸张是为了上色的位置一致，有利于后面的第二次、第三次套色。飞印一般用于单色印刷，不需要固定纸张。大卫，你现在把这块版拿下来，放第二块版，印第二张。

小　　龙：这样看来，如果熟练的话，效率会特别高。

李馆长：是的。尽管之前的工序需要大量的人工和材料，可一旦开始刷印，雕版印刷的优点就能体现出来了。

大　　卫：真高效啊！

李馆长：印得不错。等完全晾干，就可以装订成册了。

Brush Printing

> Xiaolong, David, and Mr. Li went to a printing table.

Xiaolong: Ink, paper, block… Hey, what are these two things?

Mr. Li: This one is a palm brush. We use it to apply ink to the block. The other is a palm smoother. We use it to press and brush the printing paper.

Xiaolong: Oh, they're made of palm leaves.

Mr. Li: To be precise, they're made from fabrics of palm leaves.

David: Look! The paper is fixed on one side of the table. Why?

Mr. Li: Well, actually, both the paper and the block need to be fixed before printing. The block is secured with nails to stop it from sliding, and the practice of fixing the paper allows us to turn pages without losing the order.

Xiaolong: I see. Then, why is there a gap between the paper and the printing table?

Mr. Li: You'll know the answer later when you try it yourself. The printing table is designed for convenience and efficiency. Right, let's get started. We have the blocks of Tang Poems here.

David: I'd like to print Li Bai's poems.

Xiaolong: Wang Wei is my favourite Tang poet. I'd like to print Wang Wei's poems.

Mr. Li: Alright. The first step in engraved block printing is to brush red ink for the initial proofreading. Then, brush blue ink for the second proofreading before mass printing to ensure that everything is correct. Today, we can skip the previous two steps and go straight to the last step of printing.

David: OK, I'll brush the ink on the block now.

Mr. Li: No rush. Before that, you need to brush it with some water twice.

Xiaolong: It's clean, isn't it? Why should we brush it with water again?

Mr. Li: Wetting the board with water makes it easier to apply the ink.

David: Got it. Can I brush it with ink now?

Mr. Li: Yes, you can. But there are still some techniques to consider. Look at this palm brush. Notice anything special about it?

Xiaolong: It seems to have four layers from the inside to the outside, but they're not aligned.

Mr. Li: Good, exactly. It's more intricate than it appears. The design, refined over time, allows for a continuous supply of ink without smudging the characters to be printed, so it's also called an ink equaliser.

David: I didn't realise there was so much scientific knowledge involved in this simple brush.

Mr. Li: You see, ancient Chinese people were really clever. Now, let's soak the brush in warm water before applying the ink.

David: Why?

Mr. Li: Touch it when you take it out of the water, and you'll see.

David: Oh, the brush becomes very soft. It's much easier to apply ink this way.

Mr. Li: Exactly. Now, you can move the brush in circles in the ink tray to get it evenly coated, and then apply the ink to the block in circular motions. You should work quickly to ensure the ink is evenly distributed and the colour of the print is consistent in shade.

David: Can I place the paper on the block now?

Mr. Li: Yes. It's called paper attaching. Watch me. Hold the paper like this, lay it flat on the block, and use a clean palm smoother to press the paper firmly.

Xiaolong: Look! The characters are appearing.

Mr. Li: We should do it slowly. See here, the characters are a bit vague; you can press harder to make them more distinct. Now you can take the paper off the block.

> In this way, the first page of the Tang poems was printed on the paper. The print was fresh and clear with semi-wet ink.

David: Where should I put this page? The ink seems to be still wet.

Mr. Li: Now, you see, this gap is designed for drying pages.

David: Wow, I got it! It allows the paper to fall and dry naturally. What a clever design!

Mr. Li: Definitely so. But it's not for drying only. The method we're using now is press-printing. Since the paper is fixed on the table, we need a place to collect the paper after printing.

David: Oh, it's related to the technique.

Xiaolong: Indeed. If the paper isn't fixed on the table, we wouldn't need that gap, right?

Mr. Li: Exactly. There are two types of printing methods. One is press-printing, typically used for overprinting. It's important to fix the paper like this to ensure the colours align correctly. The other is called fly-printing, where it's not necessary to fix the paper. David, now take this block off, put in the second one, and continue the printing.

Xiaolong: The printing can be very fast once you are skilled.

Mr. Li: Yes. Once the printing starts, it becomes efficient and highly productive, even though the earlier steps require lots of labour and materials.

David: It's really efficient!

Mr. Li: Well done! Now just let them dry, and then we can bind the pages into a book.

线装

> 小龙和大卫捧着自己亲手印制的唐诗,来到体验馆的最后一个台子。台子上有纸捻、线团、针和打孔器,旁边还有一些蓝色硬纸。

小　　龙：李馆长,这儿有订书机吗?我想把这些唐诗装订起来。

李馆长：雕版印刷的东西可不是用订书机装订的,得好好装帧一下。看到工具了吗?

大　　卫：是要用针线把它们装订起来吗?我见过线装书。

李馆长：是啊,书页印出来后要装订到一起。雕版印刷兴起后,书的装帧方式也在不断改进。从蝴蝶装、包背装一直改进到线装。

大　　卫：装订方式居然也有这么多种。我们能自己试试吗?

李馆长：当然可以。要注意,古时候的书写方式是从右往左、从上到下。你们用印好的李白诗和王维诗来装订吧。来,先试试最早的蝴蝶装。

蝴蝶装　Butterfly Binding

小　　龙：蝴蝶装？这个名字好浪漫呀。

李馆长：你们看，打开书页后，这版心像是蝴蝶身躯，两侧的书页好像翅膀，所以叫蝴蝶装。你们再看一下印的书页，是不是一面有字、一面没字？

小　　龙：对。刚才师傅说这是单面印刷。现在我们可以装订了吗？

李馆长：好的。我们先装订李白的诗吧。先以有字的一面为准，字对字对折，再把所有折好的书页按顺序整理好。

大　　卫：没问题。我来装订吧。李白的诗浪漫，用蝴蝶装很合适。

李馆长：好，你把浆糊刷在折痕处，再用一张硬纸把书裹上，作为封底和封面。

大　卫：怎么不用针线呢？

李馆长：蝴蝶装是不用针线的。小龙，打开看看。

小　龙：不对啊。第一页和第二页中间是两张空白页。还有另外一个问题，只用浆糊粘能行吗？要是翻的次数多了，书肯定会散开的。

李馆长：你说得没错。正是因为蝴蝶装使用体验不好，长期翻阅又容易散落。所以，后来出现了改进的包背装。

大　卫：包背装怎么做？

李馆长：和蝴蝶装相反，包背装是以空白一面为准，把纸对折。

小　龙：好的。我来试试。

包背装　Back-Wrapped Binding

李馆长：要把所有折好的书页按顺序整理好。在书页开口处打两个孔，用纸捻固定。最后，在开口处刷上浆糊，用一张硬纸把书页裹上。

大　卫：这样好，每页都有字。

李馆长：包背装确实牢固了很多，但翻多了，书页还是容易散落。改进后的线装就不容易散了。线装一开始的几道工序和包背装一模一样。

小　龙：太好了。我们知道怎么装了。大卫，我们来装订王维的诗。我来折页，你接着装订，怎么样？

大　卫：好了，王维的诗集装订好啦！接下来就该加书皮了吧？

线装　Thread Binding

李馆长：对。线装是用与书页大小相同的封面和封底，书上面一张，下面一张，粘在书上。

小　龙：好的。既然是线装，应该是用线把书缝起来吧。

李馆长：是的。缝线有很多种方法。最普遍的是四目式。就是在书页开口处打四个孔，用线装订成册。

大　卫：看来用线装最结实了。

> 参观了雕版印刷过程，并亲自装订，小龙和大卫终于弄清楚了如何印制和装订一本书。

小　龙：要是配上五颜六色的插图就更好了。

李馆长：当然可以了。雕版印刷可不只是在白纸上印黑字啊，也可以印出色彩丰富的精美图册。走，我们回雕版印刷展厅，去看看印制好的书和画。

Thread Binding

> Holding the printed pages of Tang poems, Xiaolong and David went to the last table in the workshop. On the table were paper threads, balls of threads, needles, and a hole-punch, alongside some blue hard paper.

Xiaolong: Mr. Li, could I borrow a stapler? I want to bind these pages of Tang poems together to make a small book.

Mr. Li: We usually do not staple them. Here are the binding tools. We'll bind them carefully.

David: So, are we going to bind them with a needle and some thread? I've seen thread-bound books before.

Mr. Li: Yes, we'll bind the printed pages into books. Traditional book binding in China has evolved from butterfly binding and back-wrapped binding to thread binding.

David: That's interesting. Can we give it a try?

Mr. Li: Of course. And just remember, in ancient Chinese scripts, the writing order was from right to left, and from top to bottom. Be careful when binding the book pages. You can start with the butterfly-binding method.

Xiaolong: Butterfly-binding? The name sounds very romantic, right?

Mr. Li: Yes, it's so called because the book spine resembles the body of a butterfly. And when you open it, the pages on both sides are spread out like butterfly wings. Now look at the book pages you've printed. Have you noticed that characters are printed on only one side of each page?

Xiaolong: Yes, that's true. I was told it's one-side printing. Shall we start the binding now?

Mr. Li: OK. Let's start with Li Bai's poems. Fold the side with characters and arrange all the pages in order.

David: No problem. I'll handle the binding. Li Bai's poems have a romantic touch, and butterfly-binding suits them perfectly.

Mr. Li: Good. Now, brush some paste on the folds first, and then cover the book with two pieces of thick paper as the front and back covers.

David: Why not use the thread and a needle?

Mr. Li: Well, butterfly binding doesn't need any thread or needles. Xiaolong, could you open the book and have a look?

Xiaolong: It looks like there might be something wrong? There are two blank pages between the first and second pages. Besides, will it be sturdy enough with just paste? It seems that it could easily fall apart.

Mr. Li: You're right. Due to these limitations, back-wrapped binding replaced butterfly binding, as a more reliable method.

David: How is the new method specifically carried out?

Mr. Li: Unlike butterfly binding, we can fold the pages with the blank side inside.

Xiaolong: Alright, I'll give it a try.

Mr. Li: You should arrange all the folded pages in order and make two holes in the open sides. Then secure them with thread. Next, apply some glue and wrap the pages with a piece of thick paper as a cover from the front to the back.

David: This method seems better. Every page has characters on it.

Mr. Li: It's much better, but it still tends to fall apart over time. So a further improved method, the thread-stitched binding emerged later. The initial steps of the new method are the same as those of the back-wrapped binding.

Xiaolong: Understood. David, let's bind Wang Wei's poems together? I'll handle the folding, and you can do the binding.

David: No problem. Well, I've finished. Is it time to add the cover?

Mr. Li: Yes, indeed. Take two pieces of thick paper, the same size as the pages, and paste them on the front and back respectively.

Xiaolong: Got it. Then, the book is stitched with threads, isn't it?

Mr. Li: Yes. You're right. There are various stitching methods. The most common one is four-hole stitching. You punch four holes in the open side of the book, and then bind it with the thread.

David: Wang Wei's poems should be securely bound this way.

> Xiaolong and David understood the entire process of engraved block printing, after visiting the experience centre and binding a book of poems themselves.

Xiaolong: It would be nicer to have some colourful illustrations in the book.

Mr. Li: Absolutely. In engraved block printing, not only black characters are printed on white paper, but also the colourful and beautiful patterns. Let's return to the exhibition hall to see more printed books and paintings.

套色印刷

> 李馆长和小龙、大卫一起回到雕版印刷展厅,来到套色印刷展台前。

李馆长:小龙,大卫,你们来看看这首诗,这是三色套印。

小　龙:雕版还可以印三种颜色?

李馆长:神奇吧?猜猜看,是怎么印上去的?

大　卫:应该是在印版上同时刷上不同的颜料,对不对?

李馆长:对。是在一块版上刷不同的颜料。刚刚你们印的唐诗是黑色字,那叫雕版单色印刷。这个叫雕版彩色套印。

大　卫:我有个问题,不同颜色会不会串色呀?

李馆长:会的。你们再想想,该怎么印才不串色?

小　龙:几种颜色分开印吗?

李馆长:对的。雕版彩色套印分单版和多版两种。单版套印是彩色套印的早期形态，有两种方法。一种方法是在一块版上的不同区域，涂抹上不同的色彩，只需刷印一次；另一种是分色上版，刷印一种颜色时，要将其他颜色区域遮盖起来，然后多次覆纸刷印。

小　龙：如果多次涂抹、多次覆纸的话，一定得保证不同色彩相对应的位置很准确，对吧？

李馆长：没错。但这样的方法用于印书还行。要是印画，可能就会出问题。后来套印技术不断发展，就出现了分版分色套印。

大　卫：分版？要用很多版片吧？那是不是很繁琐呀？

李馆长：是的。分版分色套印一般用于版画的印制。与单色印刷相比，分版分色套印的工序更繁琐复杂。雕版时，要根据颜色来决定做几块印版。套印时，因为要将两种以上颜色刷印在同一张纸上，还需要一个套版的过程。

小　龙：听起来很难操作呀！

李馆长：的确是，所以需要更高超的技术。第一步，要在一块版上涂刷一种颜色后覆纸刷印。第二步，在另一版上涂刷另一种颜色，再覆上第一次刷印过的纸，二次刷印。必须保证版框完全吻合，才能完成一张双色套印。如果印刷时两块版不吻合，或者刻版时两块版上的内容相对应的位置算得不准确，就会出现文字、图案参差不齐，或者相互重叠，那就出废品或次品了。

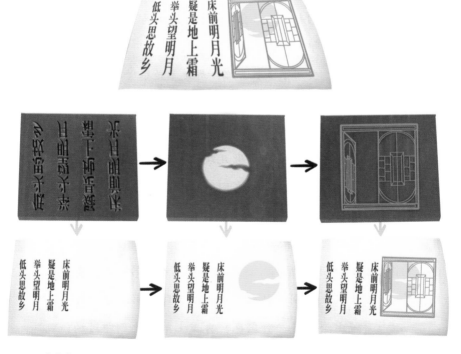

套色印刷　Overprinting

大　卫：如果颜色更多呢？是要做更多的版片吗？

李馆长：对的。套印几种颜色，就得做几种版片。多的有五色套、七色套，甚至九色套。套色越多越繁琐。

小　龙：原来分版分色是这样啊。刚才您说套印技术大多应用于版画，那版画是什么时候出现的？

李馆长：在中国，版画是与刻书一起出现的，但两者发展得不平衡。1000多年前的宋朝刻书就已经很成熟了，又过了几百年，到明朝后期，版画才发展成熟。

大　卫：为什么呢？

李馆长：这与社会需求和雕版技艺发展有关。可以说是套版印刷的完善推动了版画发展。套印起源于南宋的纸币印刷，13世纪中期，出现了朱墨两色套印书。

小　龙：古人用多种颜色刷印，是为了好看吗？

李馆长：不全是。中国古代一般把经、史、传类书籍当教科书使用。不同的学者会根据自己的研究和理解，对原文本进行批注，这样一本书就会出现很多不同的版本。为了把原文同批注区别开，古人用墨笔写原文，朱笔写批注。到了雕版印刷时期，自然就有了朱墨两色的印刷品。我们称它"朱墨套印本"或"双印本"。至于套印纸币，你们知道为什么吗？

大　卫： 那肯定是为了防伪。后来为什么很长时间没有发展呢？

李馆长： 因为套版印刷费时、费工，成本很高，所以很长时间并没有得到推广。直到16世纪末，随着技术成熟和需求增长，套色印刷才开始盛行，出现了多色套印的书。

小　龙： 颜色丰富了，就可以用来印画了吧？

李馆长： 是的，起初套印主要用于印书。后来颜色增多，技术也越来越成熟，又出现了印画的新技术，版画才随之兴盛起来。走，我们去看看木刻水印画吧。

Overprinting

> Xiaolong, David and Mr. Li went to the overprinting zone in the exhibition hall.

Mr. Li: Xiaolong, David, look at this. It's three-colour overprinting.

Xiaolong: Wow! Three colours are printed on one page. How does that work?

Mr. Li: Quite impressive, isn't it? Can you guess how it's printed?

David: I reckon the block could be painted with different colours all at once. Am I on the right track?

Mr. Li: Yes. Just now, you've printed the poems in black, known as engraved single-colour printing, and this is overprinting.

David: But I wonder if different colours will blur and overlap?

Mr. Li: Well, they will indeed. Any guess on how to prevent that?

Xiaolong: By printing colours separately?

Mr. Li: Correct. Overprinting can be divided into single-block and multi-block methods. The former can be further divided into two ways. One involves applying different colours to different parts of a block and printing all colours at the same time; the other involves dividing the block into different colour sections and printing one colour at a time while covering up other parts until all the colours are printed out.

Xiaolong: Is it tricky to keep all the colours in their own places?

Mr. Li: Indeed. It's fine for printing books, but not for printing paintings. With the advancement of the overprinting technology, the more sophisticated multi-block overprinting method was developed.

David: Multi-block? That sounds like it would require a lot of blocks, right? It seems very complicated.

Mr. Li: Absolutely. It's commonly used to print paintings. This multi-block overprinting is much more complicated. You have to consider how many blocks are needed to print all the colours on one piece of paper. A specific overprinting sequence is necessary because more than two colours are applied to the same piece.

Xiaolong: It sounds quite challenging.

Mr. Li: Indeed. That's why we need more advanced techniques. During printing, the first step is to brush the first block and place the paper on it; the second step is to brush another colour on another block, then place the already printed paper on it and print again. You have to ensure that the two blocks are in exactly the same position to achieve a perfect two-colour overprint. If the two blocks don't align correctly, or if the content positioned for engraving isn't accurately calculated, it can result in jagged texts and patterns. That would produce flawed or inferior products.

David: What if there are more colours? Does it mean we need more blocks?

Mr. Li: Yes, exactly. The number of blocks depends on the

colours needed for the final print. The more colours we have, the more blocks we use. We have five-colour, seven-colour, and even nine-colour overprinting. It gets more intricate with more colours.

Xiaolong: I see. You just mentioned that overprinting is commonly used in woodcuts. When exactly did woodcuts begin?

Mr. Li: Woodcuts and block-printed books emerged almost at the same time in China. However, they didn't progress at the same rate. Block-printed books were produced on a large scale in the Song Dynasty more than 1,000 years ago, but the woodcut didn't mature into an art form until a few hundred years later, during the late Ming Dynasty.

David: Why did they develop at different paces?

Mr. Li: It's closely related to the social needs and the progress of engraved block printing. Overprinting greatly influenced the development of the woodcut. In fact, overprinting originated from paper currency printing in the Song Dynasty. By the mid-13th century, red and black overprinting was used for book printing.

Xiaolong: I reckon multi-colour printing is meant for aesthetic effects.

Mr. Li: Not entirely. In ancient China, some Chinese classics, historical and biographical books were commonly used as teaching materials. Scholars liked to add notes and comments in the book margins. Then the books were copied with the same title but different annotations. The black ink was used for the original text, and the red ink for annotations to differentiate them. Thus, the earliest bicolour overprinting appeared in black and red. It's known as red and black printing or bicolour printing. And do you know why paper currency was overprinted in the Song Dynasty?

David: I suppose it was used to prevent counterfeiting. But I wonder why it didn't continue.

Mr. Li: Because it was time-consuming, labour-intensive, and costly, it wasn't widely adopted for long. However, by the late 16th century, with the advancements in printing technology and a growing demand for coloured works, the overprinting technique flourished in producing colour-printed works.

Xiaolong: So, paintings were printed with this technique, right?

Mr. Li: Absolutely. Early overprinting was primarily used for book printing. As time went by, woodcuts gradually progressed with improved techniques. Let's head over to the exhibition hall and take a look at them.

饾版和拱花

> 小龙和大卫跟随李馆长来到版画展区,欣赏用饾版和拱花制作的精美画作。

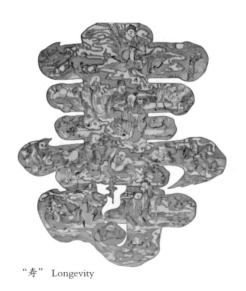

"寿" Longevity

大　　卫：你们看，这个"寿"字里面居然嵌入了很多画。"寿"字里面有这么多动物、植物和人物。

李馆长：大卫，这些动植物和人物都与祝寿有关。你看，这个脑袋突出的人是寿星，神仙、仙女也都是来祝寿的。梅花鹿、桃子代表长寿。可爱的小朋友表示子孙绵延，家族兴旺。

大　　卫：小龙给我看剪纸纹样的时候也讲过这些。

小　　龙：李馆长，这些不同的颜色是怎么印上去的？

李馆长：这个"寿"字和旁边的《十竹斋笺谱》都叫木刻水印画。

小　　龙：木刻水印？这不就是常见的中国画吗？

李馆长：木刻水印是当代北京荣宝斋[①]在饾版的基础上发展完善的，是一种专门用来复制中国画的工艺。这个技术很复杂，也非常厉害，能很好地复原原画的风格。好的木刻水印能达到乱真的程度，是"再创造的艺术"。

小　　龙：我不太明白，找人画画，为什么还要复制呢？

李馆长：这是复制古画的需要。中国的很多古画年代久远，原画很脆弱，一般不会拿出来展出。那就需要复制原画，而且还要复制得像，于是就产生了这个技术。

小　　龙：听说荣宝斋的木刻水印画能以假乱真呢。

李馆长：是的，优秀的师傅技术非常高超，用木刻水印制作的画真的能以假乱真。这里面还有故事呢。

大　卫：什么故事啊，李馆长，您快讲讲。

李馆长：大画家齐白石②多次去荣宝斋亲自指导印制自己的画。有一次荣宝斋完成了《虾》的印制，在齐白石面前挂出他的原作和印制的画。齐白石看了半天，竟然分不清哪个是自己画的。

大　卫：这真的很厉害。

小　龙：居然在画家本人面前都能以假乱真，想不到。

大　卫：李馆长，那什么是饾版呢？

李馆长：饾版印刷这种印画方法，就是在套版基础上，将彩色画稿按不同颜色分别勾摹下来，每种颜色刻成一块小木版，然后依照色调由浅入深依次套印或叠印，因其形似饾饤而得名。

大　卫：那饾饤是什么呢？

李馆长：古时候，人们将五色小饼堆叠在点心盒中，就叫"饾饤"。这种印画方式与饾饤点心摆放的样子很像，所以称为饾版。

小　龙：那现在的木刻水印具体流程有几步？

李馆长：有四步。第一步是勾描。拿到原稿后，先根据色彩及画面大小确定分成多少块印版。然后用透明的胶纸覆在原作上，把画面上的点、线、色块、文字等如实地勾描下来。再用半透明的薄纸蒙在描好的胶纸上，按照不同层次、不同颜色细致地描绘成一张张底稿。

小　　龙：临摹底稿对技术要求很高吧？

李馆长：是的。需要有一定的绘画功底。第二步，上样，将勾描好的底稿分别粘在底板上。第三步，依照原画的风格和笔法精心雕刻。

大　　卫：那第四步就是刷印了吧？

李馆长：对的。第四步，按照顺序依次套印。首先要刷水，湿润版块，之后刷墨。刷印时，用棕帚或毛笔把中国画的点、线、色块、文字等按照原样、原色，用不同的方法刷涂在不同的印版表面。然后由浅致深、由淡到浓反复刷印同一张纸，直到完成套印。当然，在实际操作中，还有很多非常细致的工序和专业技巧，我就不多说了。

小　　龙：那应该是印在宣纸上吧？

李馆长：要看具体情况，有的用宣纸，有的用绢。

大　　卫：在刷印时会用到松烟墨吗？

饾版和拱花 Block-Assembled Overprinting and Embossed Overprinting

饾版 拱花 Block-Assembled Overprinting and Embossed Overprinting

李馆长：那要看画的类型。工笔画一般使用水性植物颜料，水墨画才会用到松烟墨。

大 卫：你们看，这个小画好精致呀。咦，这个梅花的花瓣没有上色，却是立体的，好特别呀。

小 龙：大卫，你看这两张，第一张是瓶口内侧突出，第二张是水流凸显。

李馆长：这个技术叫拱花，是以图案为基准，将其轮廓线或纹理制成拱花版，用凸凹雕版嵌合压印在纸上，从而得到凹凸纹理的立体效果。饾版、拱花两种技艺结合起来，画面上会产生丰富的视觉效果。

小　　龙：拱花都是没有颜色的吗？

李馆长：不一定。拱花分有色拱花与无色拱花两种。

大　　卫：可真神奇呀。我还是第一次看到这样的画呢。

李馆长：这些精致的小幅画出自《十竹斋笺谱》。

小　　龙：我有个问题，这些都是画，为什么叫笺谱，不叫画谱呢？

李馆长：笺谱是一个很有中国文人特色的东西，原来是指印有淡雅图案的信纸，特别受文人雅士喜爱。后来，有人专门将印有图画诗文的漂亮笺纸归类整理，汇编成册，称为笺谱。现存最有名的是1644年胡正言做的《十竹斋笺谱》，一共4卷，283幅画，由饾版、拱花印制。《十竹斋笺谱》代表了古代彩印的最高水平。

小　　龙：拱花是谁发明的？这个人可真聪明呀。

李馆长：这种印画技术就是胡正言首创的。在他那里，雕版印刷从一项印刷技术发展成了个性化的艺术创作。

小　　龙：雕版印刷的画也需要装裱吗？

李馆长：是的，最后一步是把印好的画装裱起来。这样才算完成一幅画作的复制工作。接下来，我们再到版本区看看。

注释：
① 荣宝斋：北京一家老字号店铺，经营文房四宝，迄今已有300余年历史。
② 齐白石：1864—1957，近现代中国绘画大师，世界文化名人，擅画花鸟、虫鱼、山水、人物。

Block-Assembled Overprinting and Embossed Overprinting

> Xiaolong, David and Mr.Li went to the overprinting zone, enjoying the beautiful works of block-assembled overprinting and embossed overprinting.

David: Look, this character *shou* (寿) is actually embedded with many paintings. Wow, so many images of animals, plants and figures are built into the character.

Mr. Li: David, these images are all related to the concept of longevity. You see, this figure with a protruding head is the God of Longevity. And these are some other gods and fairies who come to celebrate his birthday. The deer and peaches also symbolise longevity. The cute little kids represent the hope for more offspring in the family.

David: Yes, Xiaolong actually told me about these when he

showed me the paper-cutting patterns.

Xiaolong: Mr. Li, how were these different colours printed on the same piece of paper?

Mr. Li: Take a look at this character *shou* and the *Chinese Poetry Paper by the Master of the Ten Bamboo Hall* next to it. They're both made by using the technique of woodcut watercolour printing.

Xiaolong: Woodcut watercolour printing? I thought they were ordinary Chinese paintings.

Mr. Li: It's a traditional technique of printmaking. It was developed by Rongbao Zhai[1], a time-honoured stationery and calligraphy shop in Beijing, based on block-assembled overprinting, or *douban*（饾版）in Chinese. The technique is specially used to replicate Chinese paintings, maintaining exactly the original style. The best ones look identical to the original. We call it "the art of recreation".

Xiaolong: I'm puzzled. Why did they make copies? It's not difficult to find painters.

Mr. Li: It's all because many ancient Chinese paintings are old and fragile. The originals have to be carefully preserved and are usually not displayed to the public. Therefore, it's necessary to produce some exact copies.

Xiaolong: I heard sometimes it's hard to tell a copy by Rongbao Zhai from the original. Is that true?

Mr. Li: Yes, the craftsmen could be so skilled that the copied paintings are nearly indistinguishable from the originals. Let me share a true story with you.

David: What's the story? I'd like to hear it.

Mr. Li: Qi Baishi[2], a great Chinese painter, often supervised the overprinting of his paintings in Rongbao Zhai. Once they completed the printing of his painting *Shrimps*, they hung it together with the original. Master Qi looked at them for a while, but could not tell them apart.

David: Wow, that's really impressive.

Xiaolong: Unbelievable. The painter himself can't even recognise his own work.

David: Mr. Li, what's block-assembled overprinting then?

Mr. Li: It's a kind of printing technique. Here's how it works. Sketch different parts of the painting based on what colours you're going to print. Then carve each part into a small block, and print over each block from the lightest to the darkest colour. They call it *douban* (饾版) in Chinese simply because the assembled small blocks look like *douding* (饾饤), a kind of plating technique used for displaying cookies in ancient China.

David: What's that?

Mr. Li: Back in ancient times, cookies of diverse colours are often placed in a plate, known as *douding*. This printing style resembles the way cookies are arranged in the plate, so it's called *douban* or block-assembled overprinting.

Xiaolong: Then what's the procedure for watercolour overprinting?

Mr. Li: There are altogether four steps. First, you have to draw the outline of a painting. And decide how many blocks you need for printing the painting, considering the colours and sizes of the blocks needed, and lay a special kind of transparent paper over the original. Later, copy the dots, lines, coloured parts and texts of the whole painting onto some pieces of semi-transparent paper according to different colours and layers. These are just the manuscripts for the engraving.

Xiaolong: It's technically demanding, isn't it?

Mr. Li: Yes. Some drawing skills are required. The second step is similar to sample pasting in book printing. Paste the drawings to the blocks separately. The third step is to engrave the blocks carefully to show the original style and brushwork.

David: The fourth step is brush printing, right?

Mr. Li: That's right. The fourth step is to overprint the painting in order. Before printing, brush to moisten the blocks with water, and then with ink. When printing, use a palm brush or a calligraphy brush to brush the dots, lines, colours or characters on the blocks according to the original. And then overprint the same paper on all the blocks one by one from light to dark to finish a perfect painting printing. Of course, it's a sophisticated process. We'll talk about it later.

Xiaolong: It's supposed to be printed on *xuan* paper, right?

Mr. Li: It depends. Some are printed on paper, and some on silk.

David: Do we use pine-soot ink this time?

Mr. Li: Well, it depends on the type of painting. Colour pigments are usually used for paintings of figures, flowers and birds, and pine-soot ink is used for black-and-white landscape paintings.

David: Look, this little painting is so exquisite. The petals of the plum blossom aren't coloured but make a vivid impression. It's so special.

Xiaolong: Yes, indeed. Also, the inside of the bottle and the lines of the stream water are raised.

Mr. Li: This technique is called embossed overprinting to enhance lines or textures. The pattern is embossed onto the paper by a set of convex-concave engraved blocks, thus producing a three-dimensional effect. The combination of block-assembled overprinting and embossed overprinting does create a rich visual effect.

Xiaolong: Is the embossed overprinting colourless?

Mr. Li: It depends. There are coloured and colourless ones.

David: Amazing. It's the first time for me to see a painting like this.

Mr. Li: These exquisite small pieces are paintings on letter-head pads, printed by the master of the Ten Bamboo Hall.

Xiaolong: I have a question. They're paintings. Why are they on letter heads?

Mr. Li: It was something special for ancient Chinese literati. Originally, these letter heads with elegant patterns were loved by scholars, and eventually, all these designs were grouped into books of letter-head pads. The most famous one that survived is made by Hu Zhengyan, the master of the Ten Bamboo Hall in 1644. A total of 4 volumes of 283 paintings of block-assembled overprinting and embossed overprinting. These works represent the highest level of ancient watercolour overprinting.

Xiaolong: Who invented the technique of embossed overprinting? It's really brilliant.

Mr. Li: The pioneer is Hu Zhengyan himself. The technique of embossed overprinting does help produce a three-dimensional effect. In his case, the engraved block printing was transformed from a technique into a personalised work of art.

Xiaolong: Are they usually framed?

Mr. Li: Yes, the final step is to frame it to complete the whole process of reproducing the painting. Now, let's go to the exhibition hall to see different kinds of printed books.

Notes:

1. Rongbao Zhai: A time-honoured store in Beijing, with a history of over 300 years, specialising in the set of tools for Chinese calligraphy and painting.

2. Qi Baishi: 1864–1957, a master of modern Chinese painting and a world-known cultural celebrity. He specialised in painting flowers, birds, insects, fish, landscapes and figures.

刻书

> 小龙和大卫随李馆长来到雕版印刷的版本区。两人看到展台上陈列着各种版本的书。

小　　龙：李馆长，这些都是雕版印刷的吗？看上去很不一样啊。

李馆长：都是的。我们收集的刻书时代不同，种类很多。

大　　卫：刻书？不是应该叫印书吗？

李馆长：刻书是雕版印刷书籍的通称。按规模一般分为三类：官刻、私刻和坊刻。

小　　龙：听上去，有公有私呀。

李馆长：是的，而且公私分明。

大　　卫：那什么样的书籍属于官刻呢？

李馆长：官刻当然是指官方刻印的书。官刻本的规格高，规模大，比如说历书。自古以来，制定、颁布历书就是中央政府的重要职责，历书可以有效保障民众不误农时。

刻书　Block-Printed Books

另一种官刻就是史书。中国历代王朝都会为上一个朝代修史，这是传统。

小　　龙：李馆长，您看，这里说，这套《全唐诗》是"中国雕版印刷第一书"，是说它质量好吗？

李馆长：这么说虽有点儿夸张，但这套书确实质量很好。实际上，这涉及官刻的一种形式，叫"局刻"。这是为雕版印刷某种书专门设立一个单位，完成这部作品后就解散相关人员。印诗就叫"诗局"，印书就叫"书局"。为印这套《全唐诗》，清朝康熙皇帝下令专设扬州诗局，召集全国各地雕版印刷的能工巧匠，用将近两年的时间完成了这套书的刻印工作。全书写、刻、校、印都非常精良。

大　　卫：皇帝下令刻印的书肯定是官刻，那什么是私刻呢？是自己家私下印书吗？怎么感觉像是偷偷摸摸呀。

李馆长：你这是从字面上理解。私刻是相对官刻而言，也是官方允许的。家族、寺庙、道观，甚至个人藏书爱好者，他们刻印的书统统称为私刻，也叫家刻。

小　　龙：私刻会刻印哪些书呢？

李馆长：一般寺院刻印经书和佛像，道观刻印符咒，祠堂刻印家谱和家训。还有一些有钱人、文人、收藏家也参与刻印。清朝时期，扬州经济富庶，文化繁荣，因而私刻发达。有钱的商人会不惜重金请人精心刻书，用来收藏。一些专业藏书家也会专门刻书。还有书画皆佳的文人会亲自操刀精刻，他们的作品还非常受欢迎呢。

大　卫：那最后一类是民间刻书了？

李馆长：没错，确切说，是书商以营利为目的而刻印的书。这类书通常叫坊刻。坊主聘请雕版印刷艺人，集中于书坊内刻印图书，逐步形成自家的刻印风格，或慢慢形成某个地方流派。你们要知道，坊刻在刻书发展史上发挥了重要作用呢。

小　龙：我还以为官刻是最先出现的呢。

李馆长：不是的。坊刻书出现得最早，地域分布最广、印刷量也最大，最先采用雕版印刷的就是民间书坊。

大　卫：看来是因为有市场需求。

李馆长：是的。因为有需求，现在依然还有雕版印刷，不过主要用于家谱刻印和木版年画生产。

小　龙：谢谢李馆长带我们了解雕版印刷的整个过程。

大　卫：我们收获真大呀。谢谢李馆长！

李馆长：不客气，我也很高兴能向年轻人介绍我们的雕版印刷文化。欢迎以后常来。

Block-Printed Books

> In the book display zone, Xiaolong and David saw many books of different versions.

Xiaolong: Mr. Li, are these all block-printed books? They look so different.

Mr. Li: Yes, they are. They come from various categories and eras.

David: Block-printed books?

Mr. Li: It's a general term for this type of book. They're typically categorised into three kinds on their scale: official printing, private printing, and workshop printing.

Xiaolong: So, some are done by the government and some by private owners.

Mr. Li: You're right! And the distinction is quite clear.

David: What's official printing, then?

Mr. Li: It's known for its high quality and large scale. Taking the calendar books as an example, it was an important task for the government to create and distribute the calendar books to ensure the proper timing for agricultural production. Another example is the history books. In ancient China, it was a tradition for a new dynasty to compile the history of the preceding dynasty.

Xiaolong: Look. It says here, this book of *Tang Poetry* is "the best of Chinese block-printed books". Does that mean it's of high quality?

Mr. Li: That's a bit of exaggeration, but it's indeed of high quality. This series is actually a collection of books printed by the workshop of bureau printing, a kind of official printing. For this kind of project, the government would set up a temporary workshop to engrave and print a book or a series of books, and disbanded the staff once the job was done. They had book bureaus for book printing and poetry bureaus for poetry printing. This book of *Tang Poetry* was printed by Yangzhou Poetry

Bureau under the mandate of Emperor Kangxi. Then, skilled craftsmen from across the country were brought in, and it took them nearly two years to complete. The engraving, proofreading, and printing were all done with great care.

David: So that's official printing. What about private printing? It sounds like something done on the sly.

Mr. Li: It's not quite like that. Private printing refers to the printing done by families, temples, or individual book lovers. It is also called family printing.

Xiaolong: What did they print?

Mr. Li: Usually, Buddha images and scriptures were printed by Buddhist temples, paper charms and spells by Taoist temples, and genealogies and instructions by esteemed families. Rich people, literati, and book-lovers also got involved. Yangzhou was very prosperous in the Qing Dynasty, so private printing flourished there. For example, rich salt merchants would spare no expense to have books beautifully printed for their private collection. Book lovers did the same. Some literati who were good at calligraphy and painting would even do the engraving themselves, and their works were quite popular.

David: And the last category is workshop printing, right?

Mr. Li: Correct. And those workshops printed the books mainly for making profits. The owners would hire craftsmen and gradually develop their own printing styles. Places with lots of printing workshops would create a local style. It was the workshop printing that played a crucial role in promoting the crafts of engraved block printing.

Xiaolong: I thought official printing came first.

Mr. Li: No, it was the private workshops that started printing books on a large scale. They produced huge volumes of books all over the country.

David: So, market demand was the driver.

Mr. Li: Yes, that's right. Today, we can still see engraved block printing because there's still some demand, mainly for family genealogies and New Year pictures.

Xiaolong: Mr. Li, thank you for accompanying and instructing us through the whole process.

David: We've learned so much. Thank you, Mr. Li.

Mr. Li: You're welcome. It's also my pleasure to share the knowledge of engraved block printing with young people.

套色版画

> 参观了扬州中国雕版印刷博物馆，小龙和大卫又兴致勃勃地去苏州桃花坞木刻年画社学习木刻年画。王社长热情地接待了他们。

小　　龙：社长好，我是小龙，这是我的朋友大卫。张教授介绍我和大卫来学习木刻年画知识。

社　　长：欢迎，欢迎！

小　　龙：这些版片好精细啊！

社　　长：你很内行啊，还知道版片呢。

小　　龙：我是现学现卖。我们去扬州参观过中国雕版印刷博物馆。这些线条这么细，图案这么复杂，画在纸上都很难呀，居然还能刻得这么精致。

社　　长：我们这里的人都有绘画功底。不过要刻好，还是得刻苦练习。

门神　Door Gods

大　卫：社长，我们可以自己动手体验一下印年画吗？

社　长：没问题，你们可以自己试一下年画的套印，不过得有耐心啊。你们看，这是什么？

小　龙：应该是大门上贴的"门神"吧。这也算年画吗？

社　长：当然。年画最早的源头应该就是门神画。门神画起源于刻画，刻画是刻书的一种衍生品，刻的是戏曲、小说中的插画。因为插画表现力很强，后来就慢慢独立出来，形成了一种特殊的中国民间装饰艺术，图案也变得越来越丰富。年画一般是用木刻水印制作。

小　龙：我知道桃花坞历史悠久，是中国南派年画的代表。

大　卫：那年画的内容是固定的，对吧？

社　长：基本上是民间喜闻乐见的喜庆事儿，有祈福迎祥、时事风俗，还有戏曲故事等等。我们桃花坞年画刻工精致，曾经在江南地区很流行，还传到过日本、欧洲呢。

小　龙：那是享誉中外了。

社　长：是啊，桃花坞年画是我们苏州人的骄傲。兴盛的时候每年生产上百万张年画呢。

大　卫：这么厉害。那是什么时候？

社　长：300年前。

小　龙：300年前就能生产上百万张，那规模还是很大的。

社　长：是啊。我们先了解一下年画的制作工序。然后，你们可以自己动手印年画。

大　卫：太好了！工序是不是很复杂？

社　长：不是很复杂。年画一般有四道工序：画稿、刻线版、刻套色板和套印。每道工序都需要认真仔细。

套色版画《一团和气》　　Overprinted Woodcut of *Complete Amiability*

小　龙：你们也用拳刀刻版吗？

社　长：是的，拳刀是我们主要的雕刻工具。版材用的是梨木。

大　卫：那我们印什么呢？

社　长：就印刚刚刻好的经典年画《一团和气》吧。这个年画的画稿出自550多年前的皇宫。

套色版画　Overprinted Woodcuts　151

《一团和气》年画　New Year Picture of *Complete Amiability*

《一团和气》印色版　Colour Blocks of *Complete Amiability*

小　龙：没想到这是一幅很古老的画。是不是皇帝希望他的子民生活幸福、一团和气呀？

社　长：应该是这个意思吧。

大　卫：这个年画的颜色好丰富呀。

社　长：对，年画色彩丰富艳丽，非常符合民间追求喜庆热闹的审美观。你们看，这是做好的线版和套色版。

大　卫：刻得真精致，这些细细的线很容易刻坏吧。

社　长：这个需要有耐心，不断练习，刻版技术熟练之前往往会刻坏很多块板子，还容易伤着手。

小　龙：看来刻版是最关键的步骤。

社　长：是很重要。我们这个行业还有"三分刻，七分印"一说。

大　卫：那就是说，年画的印刷更重要了。

小　龙：我的理解是，如果只是刻得好还不行，印不好，那也是事倍功半。

社　长：说得很对。那我们就从刻线版说起吧。和单色雕版印刷的步骤一样，刻线版需要先画草图，然后上样，最终刻成印版。套色版需要根据年画的色彩决定做几个印色版。你数数，《一团和气》上面有几种颜色？

大　卫：黑、红、橙、绿、蓝，5种，对不对？

社　长：对。这是主色调。木版年画的颜色是由不同颜色的套色版按顺序分次印上去的。每个颜色套印之后，需要将画取下晾干，然后再印下一道。所以，如果年画上有 5 种颜色，那么至少要刻 6 块版。一般年画都是 6 块版。

小　龙：还很复杂呢！

社　长：是的。我们这里最复杂的年画需要 26 块套版呢。那我们开始印《一团和气》吧。第一步先印线版。现在把线版固定好。纸张已经在刷印台上固定好了，直接捻过来一张纸就行。

大　卫：好了，我的线版印好了。接下来是套色吧。可是，该先印哪一种颜色呢？

社　长：一般来说，按照"由浅入深，由淡至浓"的顺序来。

大　卫：那就应该先印红色。可是怎么才能保证色彩跟线版完全一致呢？

社　长：你看，纸已经固定了，色版大小是一样的，上面图案的相对位置应该对得上。现在要保证每个色版在刷印台上的位置完全重合。

大　卫：我再看一下。我的红色套版的位置没问题。

社　长：那就大胆印吧。刷印色版的原则是每次上色宁少勿多。同一种颜色可以多次上色。这样少量多次，才能得到最好的效果。具体是这样的，每一块套版，想要套色达到理想效果，要反复填套，少则3次，多则7次、8次。连续填套时，必须让套色版纹丝不动。如果前一次和后一次套色发生位移，会造成重印和偏色。

小　龙：制作年画还真不容易呢。所有的年画都这样套色吗？

社　长：那倒不是。《一团和气》是分版分色套印。

小　龙：年画还有其他的印刷方法吗？

社　长：有啊。年画中的套色可分为三种。一种是分版分色套印。第二种叫馆版，不用主版，用各块套版互相配合、补充。第三种是混合套印，就是这两种方法一起用。至于选哪种方式，要看具体情况。

小　龙：社长，谢谢您。今天学到了很多，还印了精美的年画。我要去装裱一下，回家挂在客厅里。

大　卫：我要把年画寄回英国，送给家人。

社　长：祝你们和家人和和气气，团团圆圆。

小　龙：谢谢社长！再见。

大　卫：非常感谢。社长再见！

Overprinted Woodcuts

> After visiting China Engraved Block Printing Museum in Yangzhou, Xiaolong and David headed over to the Taohuawu Woodcut New Year Picture Society in Suzhou to learn more about the New Year pictures, where they were given a warm welcome by Mr. Wang, the society president.

Xiaolong: Good morning, Mr. Wang. I'm Xiaolong. This is my friend David. Prof. Zhang suggested that we come and learn more about woodcut New Year pictures.

Mr. Wang: Welcome!

Xiaolong: Look, what stunning woodcut blocks!

Mr. Wang: You know the term! You're an expert.

Xiaolong: Well, I just picked it up in Yangzhou when visiting China Engraved Block Printing Museum there. Look at these lines. They're so fine, and the patterns so intricate. It's incredible how detailed they're on wood, considering it's tough to draw even on paper.

Mr. Wang: All our staff here are painters. They have to practice loads to get good at engraving.

David: Mr. Wang, could we have a go at printing a New Year picture ourselves?

Mr. Wang: Sure thing. You can give it a try. Just be patient. Do you know what these are?

Xiaolong: I reckon these are door gods, right? Do you call them New Year pictures?

Mr. Wang: Yes. The first New Year pictures were door gods. They were made with the same technique as block-printed books. Originally, they were illustrations in opera scripts and novels, but they gradually evolved into a stand-alone art form with loads of different patterns because they were so expressive and engaging. New Year pictures are usually done with watercolour overprinting.

Xiaolong: From what I know, Taohuawu has a long history and is very famous for its New Year pictures in Southern

China.

David: So, the themes of patterns are relatively fixed, aren't they?

Mr. Wang: Yes, that's right. They're all about folklore, with themes like prayers for good luck, different customs, and ancestor worship. You also get opera scenes and story plots. The Taohuawu New Year pictures were beautifully carved, sold like hotcakes along the Yangtze River, and were even exported to Japan and Europe.

Xiaolong: It's quite popular in and outside China.

Mr. Wang: Yes, definitely. It's Suzhou's pride and joy. Millions of New Year pictures were produced per year in its heyday.

David: That was brilliant. When was that?

Mr. Wang: About 300 years ago.

Xiaolong: Wow! Millions of copies were produced 300 years ago. That's really massive!

Mr. Wang: Right. Let's get familiar with the procedure first, and then you can print pictures yourselves.

David: Awesome. Is it very complicated?

Mr. Wang: Not really, there are four steps: drawing, line-block engraving, colour-block engraving, and overprinting. You have to be very careful at each step.

Xiaolong: Do we need a fist knife?

Mr. Wang: Yes, it's our main tool. And we use pear wood for the engraving blocks.

David: So, what are we printing today?

Mr. Wang: The *Complete Amiability*, a classic New Year picture. The original was painted by a royal artist over 550 years ago.

Xiaolong: It has such a long history. I suppose the emperor wanted to inspire people to live harmoniously and happily.

Mr. Wang: I agree.

David: The colours are so brilliant.

Mr. Wang: Yes, New Year pictures are gorgeously colourful, because folk art aims for a happy and festive atmosphere. Here are the finished line blocks and colour blocks.

David: They're so delicate. It must be really hard to carve such fine lines.

Mr. Wang: Right. It takes great patience and much practice. You have to carve a lot of boards to get a good one, and it's easy to hurt your hands while doing it.

Xiaolong: So, engraving is the key step, then?

Mr. Wang: Yes, it is. But everyone in the trade knows that engraving is only thirty percent of the success; printing makes up the other seventy percent.

David: Got it. So, printing is more important.

Xiaolong: That's to say, even if the carvings are great, it would be a failure if the printing doesn't come out right.

Mr. Wang: Exactly. Let's start with the engraving. It's like single-colour printing. Draw the whole picture on paper, and then stick it on a block before carving out the lines. For overprinting, you have to figure out how many colour blocks you need based on the original picture. Could you tell me how many colours are there in this picture?

David: Black, red, orange, green, blue. So, five colours, right?

Mr. Wang: Good. They're main colours. The picture should be printed with separate colour blocks in sequence. After printing each colour, the paper needs to be set aside to dry. This step is repeated for each colour. At least six blocks should be engraved for a picture with five colours. That's quite normal for printing a New Year picture.

Xiaolong: That sounds pretty complicated.

Mr. Wang: Yes, the most complicated picture we have needs 26 blocks. Let's print the picture *Complete Amiability*. Now, find the line block and fix it to the table. See, the paper is already there. Pull a piece and print.

David: Alright, I've finished. So, I'll do the overprinting next. Which colour should I start with?

Mr. Wang: Usually, we start from the light colour and move on to the darker one.

David: It should be red then. How do I make sure the lines and colours match up properly?

Mr. Wang: Look, the paper is already fixed in place. We've got five colour blocks, all the same size. Just make sure each one is positioned exactly right when you print.

David: Let me check... My red block is in the right spot.

Mr. Wang: Go ahead and print then. Don't apply too much paint in one go. You'll need to print the same colour multiple times to get the best result. Usually, you need to apply the colour at least three times, but for the finest quality, you can do it for seven or eight times. Make sure the block stays exactly in the same position each time, or you'll end up with a blurry or off-coloured image.

Xiaolong: It's not easy to make these, is it? Are they all done the same way?

Mr. Wang: Not all of them. This one is overprinted using separate blocks and colours.

Xiaolong: Are there any other ways to print them?

Mr. Wang: Yes, there are three methods. One is multi-block overprinting, another is block-assembled overprinting, and the third is a mix of both. The method chosen depends on the type of pictures you're printing.

Xiaolong: Thank you, Mr. Wang. I've learned a lot today and printed a beautiful picture. I'll frame it and hang it in my living room.

David: I'm going to post mine back to England for my family.

Mr. Wang: I wish you and your family peace and happiness.

Xiaolong: Thank you, Mr. Wang. Goodbye.

David: Many thanks! Goodbye.

结束语

雕版印刷术是中华文明浓墨重彩的一笔。它起于隋唐,兴于宋元,盛于明清。其产生的基础是中国大一统的国家制度和汉字的特殊性。文官科举制度的建立、民间对历书等实用书籍的需求,以及宗教传播等因素,促进了雕版印刷的兴盛。雕版印刷技艺在亚欧地区广泛传播,影响深远,对现代印刷业以及中国文化的传承与发展做出了重要贡献。

Summary

As an item of intangible cultural heritage, engraved block printing holds a significant position in Chinese civilization. With its root in the Tang Dynasty, the craft of engraved block printing was developed in the Song and Yuan dynasties and flourished greatly during the Ming and Qing dynasties. Its birth can largely be attributed to the unity of the nation and its writing system. It gained popularity because of the implementation of the imperial examination system, the increasing demand of literacy and practical knowledge among ordinary people, and the rapid spread of various religious beliefs. The technique of engraved block printing had once been introduced to many areas and countries in both Asia and Europe and had great impact on the development of modern printing technology. Its significant role in preserving and disseminating Chinese culture can never be overstated.

中国历史纪年简表
A Brief Chronology of Chinese History

夏	Xia Dynasty			c. 2070—1600 B.C.
商	Shang Dynasty			1600—1046 B.C.
周	Zhou Dynasty	西周	Western Zhou Dynasty	1046—771 B.C.
		东周	Eastern Zhou Dynasty	770—256 B.C.
		春秋	Spring and Autumn Period	770—476 B.C.
		战国	Warring States Period	475—221 B.C.
秦	Qin Dynasty			221—206 B.C.
汉	Han Dynasty	西汉	Western Han Dynasty	206 B.C.—25
		东汉	Eastern Han Dynasty	25—220
三国	Three Kingdoms			220—280
西晋	Western Jin Dynasty			265—317
东晋	Eastern Jin Dynasty			317—420
南北朝	Northern and Southern Dynasties	南朝	Southern Dynasties	420—589
		北朝	Northern Dynasties	386—581
隋	Sui Dynasty			581—618
唐	Tang Dynasty			618—907
五代	Five Dynasties			907—960
宋	Song Dynasty			960—1279
辽	Liao Dynasty			907—1125
金	Jin Dynasty			1115—1234
元	Yuan Dynasty			1206—1368
明	Ming Dynasty			1368—1644
清	Qing Dynasty			1616—1911
中华民国	Republic of China			1912—1949
中华人民共和国	People's Republic of China			1949—

图书在版编目（CIP）数据

中国雕版印刷技艺：汉英对照/刘韶方，单旭光主编．－－南京：南京大学出版社，2024.8
（中国世界级非遗文化悦读系列/魏向清，刘润泽主编．寻语识遗）
ISBN 978-7-305-26444-3

Ⅰ.①中… Ⅱ.①刘…②单… Ⅲ.①木版水印－介绍－中国－汉、英 Ⅳ.① TS872

中国版本图书馆 CIP 数据核字（2022）第 248923 号

出版发行	南京大学出版社
社　　址	南京市汉口路 22 号　　　　邮　编　210093
丛 书 名	中国世界级非遗文化悦读系列·寻语识遗
丛书主编	魏向清　刘润泽
书　　名	**中国雕版印刷技艺：汉英对照**
	ZHONGGUO DIAOBAN YINSHUA JIYI: HANYING DUIZHAO
主　　编	刘韶方　单旭光
责任编辑	张淑文　　　编辑热线　（025）83592401
照　　排	南京新华丰制版有限公司
印　　刷	南京凯德印刷有限公司
开　　本	880mm×1230mm　1/32 开　　印张 5.75　　字数 119 千
版　　次	2024 年 8 月第 1 版　2024 年 8 月第 1 次印刷
ISBN 978-7-305-26444-3	
定　　价	69.00 元

网址：http://www.njupco.com
官方微博：http://weibo.com/njupco
官方微信号：njupress
销售咨询热线：（025）83594756

* 版权所有，侵权必究
* 凡购买南大版图书，如有印装质量问题，请与所购图书销售部门联系调换